Smart Cities in Poland

This book considers and examines the concept of a Smart City in the context of improving the quality of life and sustainable development in Central and Eastern European cities.

The Smart City concept has been gaining popularity in recent years, with supporters considering it to be an effective tool to improve the quality of life of the city's residents. In turn, opponents argue that it is a source of imbalance and claim that it escalates the problems of social and economic exclusion. This book, therefore, assesses the quality of life and its unsustainability in Central and Eastern European cities within the context of the Smart City concept and from the perspective of key areas of sustainable development. Using case studies of selected cities in Central and Eastern Europe and representative surveys of Polish cities, this book illustrates the process of creating smart cities and their impact on improving the quality of life of citizens. Specifically, this book investigates the conditions that a Smart City has to meet to become sustainable, how the Smart City concept can support the improvement of the residents' quality of life and how Central and Eastern European countries create smart city solutions.

Containing both theoretical and practical content, this book will be of relevance to researchers and students interested in smart cities and urban planning, as well as city authorities and city stakeholders who are planning to implement the Smart City concept.

Izabela Jonek-Kowalska is a full professor in the Department of Economics and Computer Sciences at the Silesian University of Technology (Poland). For over 20 years, she has been dealing with the economics of business entities, including city management and finances. Her output includes over 300 publications at national and international publishers, as well as many undertakings and projects implemented in cooperation with the business environment. Her scientific interests include economics of business entities, business finance, Smart City, mineral economics, risk management and value management.

Radosław Wolniak is a full professor in the Department of Economics and Computer Sciences at the Silesian University of Technology (Poland). For over 20 years, he has been dealing with the issues of quality management, including the quality of life. His achievements include over 500 publications issued at national and international publishers, as well as many undertakings and projects implemented in cooperation with the business environment. His scientific interests include quality management, quality of life, Smart City, Industry 4.0, methods of quality management and the problems of people with disability in public management.

Routledge Research in Planning and Urban Design

Routledge Research in Planning and Urban Design is a series of academic monographs for scholars working in these disciplines and the overlaps between them. Building on Routledge's history of academic rigour and cutting-edge research, the series contributes to the rapidly expanding literature in all areas of planning and urban design.

Culture and Sustainable Development in the City
Urban Spaces of Possibilities
Sacha Kagan

Shrinking Cities in Reunified East Germany
Agim Kërçuku

Lost Informal Housing in Istanbul
Globalization at the Expense of Urban Culture
F. Yurdanur Dulgeroglu-Yuksel

China's Railway Transformation
History, Culture Changes and Urban Development
Junjie Xi and Paco Mejias Villatoro

City-making, Space and Spirituality
A Community-Based Urban Praxis with Reflections from South Africa
Stéphan de Beer

Smart Cities in Poland
Towards sustainability and a better quality of life?
Izabela Jonek-Kowalska and Radosław Wolniak

For more information about this series, please visit: www.routledge.com/
Routledge-Research-in-Planning-and-Urban-Design/book-series/RRPUD

Smart Cities in Poland

Towards sustainability and a better quality of life?

Izabela Jonek-Kowalska and
Radosław Wolniak

Routledge
Taylor & Francis Group
LONDON AND NEW YORK

Designed cover image: © Shutterstock

First published 2024
by Routledge
4 Park Square, Milton Park, Abingdon, Oxon OX14 4RN

and by Routledge
605 Third Avenue, New York, NY 10158

Routledge is an imprint of the Taylor & Francis Group, an informa business

© 2024 Izabela Jonek-Kowalska and Radosław Wolniak

The right of Izabela Jonek-Kowalska and Radosław Wolniak to be
identified as authors of this work has been asserted in accordance with
sections 77 and 78 of the Copyright, Designs and Patents Act 1988.

British Library Cataloguing-in-Publication Data
A catalogue record for this book is available from the British Library

ISBN: 9781032412481 (hbk)
ISBN: 9781032414621 (pbk)
ISBN: 9781003358190 (ebk)

DOI: 10.4324/9781003358190

Typeset in Galliard
by codeMantra

The Open Access version of Chapters 2 and 6 are funded by Silesian
University of Technology.

Contents

Foreword

The Smart City concept has been gaining popularity in recent years. It develops dynamically both in the field of social and technical sciences. Supporters of its implementation consider it an effective tool to improve the quality of life of the residents. In turn, opponents argue that it is a source of imbalance and it escalates the problems of social and economic exclusion. Bearing in mind the above circumstances, the authors of the monograph attempt to comprehensively consider and examine the concept of a Smart City in the context of improving the quality of life and sustainable development in Polish cities. The originality of the proposed research topic results from the following circumstances:

- focusing research problems around issues related to the quality of urban life, which is discussed much less often in literature and research;
- entering the quality of life assessment in the areas of sustainable development (technological, economic, social, environmental);
- analyzing the conditions of smart city development in Central and Eastern Europe (they are not the subject of frequent consideration and research);
- conducting research on a representative research sample of 280 Polish cities, allowing to verify in practice many opinions and views related to urban life, including those concerning unsustainability, and to reserve the Smart City concept only for large or financially wealthy cities;
- identifying the real and desired relations between elements of the triad determining directions of smart city development (sustainable development – smart city – quality of life);
- developing a model of pro-quality management of a smart sustainable city.

In light of the above justification, the main scientific goal of the monograph is to assess the quality of life and its unsustainability in Central and Eastern Europe cities in the context of the Smart City concept and from the perspective of key areas of Sustainable Development. In the course of considerations and research – in relation to the research gap identified above – answers are also sought to the following research problems:

- What conditions does a Smart City have to meet to be a sustainable city?
- How can the Smart City concept support the improvement of the residents' quality of life?
- How do Central and Eastern European countries create smart city solutions to improve the quality of life and what problems do they face in this respect?
- What is the scale of unsustainability of the Polish cities in the economic, technological, social and environmental area?
- Whether and to what extent the quality of life in Polish cities is influenced by:
  a their economic and financial situation;
  b size expressed in the number of residents;
  c regional geographic location defined by province.

In order to solve the formulated research problems, the authors of the monograph used the following research methods:

1 in-depth literary studies in the field of Smart City, the quality of life in cities and the sustainability of smart cities;
2 case studies on selected cities in CEE countries illustrating the process of creating smart cities and their impact on improving the quality of life of citizens;
3 survey research conducted on a representative sample of 280 Polish cities from 16 provinces (50 questions diagnosing the quality of life of residents in a five-point Likert scale in 4 areas of sustainability: technological, economic, social and environmental);
4 statistical analysis of survey results with the use of descriptive statistics and tests diagnosing the dependence of the answers obtained on economic and financial situation, size and geographical location of the cities studied.

The monograph consists of four parts: theoretical, methodological, empirical and conceptual. The first part consists of three chapters (Chapters 1–3) devoted to the main issues described in the monograph. They are, respectively, the concept of a smart city, quality of life and city management. All referred to aspects are described in the context of designing and implementing smart city solutions based on literature studies. These chapters provide a basis for identifying the existing research gap and formulating a research methodology.

The second part of the monograph includes two methodological chapters (Chapters 4 and 5). Chapter 4 presents conditions for the development of smart cities in Central and Eastern Europe – the place where the research was conducted. Particular attention was paid to Polish cities, where surveys were carried out on a representative sample of 280 cities. Chapter 5 formulates the key problems and objectives of the study and describes the research methods used (case study and surveys).

The third part of the monograph, which includes four subsequent chapters (Chapters 6–9), presents the results of empirical research. These chapters have a common, unified design reflecting the three levels of describing results. The first section of each chapter deals with the analysis of determinants of urban quality of life in four key areas of sustainable development (economic, infrastructural, social and environmental). They make use of literature studies and present the results of previous research focused on the above-mentioned issues. The second section of each chapter includes a statistical analysis of 16 Polish voivodeship cities. It shows the situation of the examined cities in a geographical perspective. The third section of each empirical chapter contains the results of surveys conducted in 280 cities in Poland. Their content – like in the previous chapters – refers to subsequent areas of sustainable development.

The fourth – conceptual – part of the monograph contains two chapters (Chapters 10 and 11). Chapter 10 presents a summary of the research results and recommendations for the municipal authorities aimed at designing and implementing smart urban solutions that improve the quality of life of the inhabitants. Chapter 11 describes the actual and desired relationships in the urban triad: sustainable development – smart city – quality of life. This part is a theoretical contribution to the development of knowledge on the management of smart cities.

As the monograph contains both theoretical and practical content, it can be interesting for the academic community and the socio-economic environment, in particular for city authorities and city stakeholders (economic partners of cities, social and environmental organizations or the residents). In particular, it may be useful for city authorities that implement the Smart City concept and/or methods of measuring the quality of life of residents in the city.

The monograph can also be used as a teaching aid for students of social and technological sciences, in the field of subjects like economics of sustainable development, sustainable consumption and production, quality of life management, technical aspects of a Smart City, economic aspects of a Smart City, urban planning and sociology of the city.

It is also worth adding that in Poland and in European countries with similar socio-economic parameters (Slovakia, the Czech Republic, Hungary, Croatia), an increased interest in smart urban solutions has been observed in recent years, which bodes well for the market placement of the monograph. Urban stakeholders in other emerging or developing countries can certainly benefit from the conclusions included in the book. In addition, its content should be valuable for urban stakeholders and academia in highly developed countries, because they draw attention to the problems of regional imbalance and the need to eliminate them.

1 The genesis, essence and development of a Smart City

1.1 Creation and assumptions of the Smart City concept

Cities must constantly evolve and adapt to the changing needs of their residents. Historically, they have been bastions of innovation and supported the development of entire countries (Florida, 2014). Demographic boom and dynamic urbanization, reinforced by globalization processes and unprecedented flows of population, capital and information, are causing cities to face unprecedented challenges and need new development strategies (Lombardi et al., 2012). Accordingly, we are witnessing very dynamic socioeconomic and technological transformations, in which cities play a key role (Winters, 2011), which is why the concept of the so-called Smart Cities has become enormously popular over the past few years.

Smart Cities are receiving increasing attention from the media, technology companies and entrepreneurs, as well as local authorities and civil society. The implementation of the Smart City concept, therefore, has the potential to make a growing number of cities around the world more efficient, more technologically advanced, and by doing so, improve the quality of life for residents (The Business, 2022).

The word 'smart' has also established itself in modern scientific discourse, becoming a very fashionable adjective to describe things and phenomena that are modern, intuitive and clever. It is no different with the concept of a Smart City, which is supposed to make life in cities better, more efficient, environmentally friendly or easier for people, as will be discussed below. There is currently no single, universally accepted and valid definition of the concept. It is still a relatively new concept and it is difficult to date its origin. It encompasses such a broad and dynamic spectrum of elements that the definition could be continuously updated (Merli and Bobollo, 2014; Jonek-Kowalska and Wolniak, 2021a, 2021b, 2022). The concept is democratized, and different institutions and rankings distribute the accents somewhat differently in the process of defining it (Falco

DOI: 10.4324/9781003358190-1

et al., 2019). This creates a certain conceptual chaos, which makes it difficult to assimilate and unambiguously interpret the term. Nevertheless, the openness of the terminology gives researchers, politicians, business people and citizens real opportunities to influence how cities and communities function (Ramirez Lopez and Grijalba Castro, 2021).

The literature increasingly accepts six basic dimensions for defining a Smart City and organizing its components. These are summarized in Table 1.1 and Figure 1.1. They are smart economy, smart mobility, smart environment, smart people, smart living and smart governance. The eight components of a Smart City are smart infrastructure, smart building, smart transportation, smart energy, smart healthcare, smart technology, smart governance, smart education and smart citizen (Mohanty et al., 2016).

Table 1.1 Dimensions defining a Smart City

Category	Characteristic
Smart economy	Using innovative and flexible solutions to increase the efficiency and productivity of local economies. The primary indicators are innovation and entrepreneurship (number of startups, R&D spending, etc.), productivity, globalism (international events or exports), labor market flexibility and transformative capacity.
Smart mobility	Use of creativity or advanced technologies to manage transportation and communication (including digital). Indicators include transportation efficiency, the use of sustainable solutions, the use of public transportation, local and global transportation accessibility or technology infrastructure, such as access to Smart City cards.
Smart environment	A responsible and sustainable approach to the environment and the use of energy resources. Basic indicators are, for example, attractiveness of natural conditions, environmental protection, sustainable management of raw materials, pollution, smart buildings, use of clean energy, water consumption, clean air or the amount of green space in the city.
Smart people	A key factor of human capital, putting the educated person at the center, also understood as a catalyst for positive change. Key indicators include education (number of students and pupils, willingness to upgrade skills), creativity of individuals (number and role of immigrants, work in creative service industry, involvement in public life) or 'technological inclusion' of citizens (e.g., participation of smartphones or high-speed Internet).
Smart living	Factor highlighting the quality of life in the city through universal access to public services in the broadest sense (access to housing, culture, entertainment), security (e.g., use of technology to prevent crime), health (e.g., life expectancy and quality of healthcare), attractiveness of the place or investment in improving the quality of life of residents.

(*Continued*)

Table 1.1 (Continued)

Category	Characteristic
Smart governance	A multilevel system of city management, relying on procedures for the participation of local authorities and citizens, such as through the use of e-services, the inclusion of a civic budget, the construction of pro-citizen infrastructure or open data networks. Local development strategies should also be included in this section.
Smart energy	The backbone of the smart energy system is the smart grid. In its formal definition, a smart grid effectively integrates the activities and behaviors of all connected users, such as consumers, generators and users, who are both consumers and generators. Smart grids provide efficient, cost-effective and sustainable energy systems with low losses, higher quality of supply, system and user safety, security of supply and improved system tolerance. The smart grid enables the integration of different energy sources, from fossil fuel-based thermal energy to green photovoltaic and wind power. Future smart grids will be much more complex than their current generation. For example, the day may come when every user will also generate solar energy, biofuel energy and even wind energy. The smart grid will effectively synchronize this energy from different sources and deliver electricity at a certain voltage and frequency without any fluctuations.
Smart healthcare	Smart healthcare can be conceptualized as an amalgamation of various entities, including, but not limited to, traditional healthcare, smart biosensors, wearable devices, information and communications technology (ICT), as well as smart ambulances. The various components of smart healthcare include on-body sensors, smart hospitals and smart emergency response. Smart hospitals use a variety of mechanisms for their operation, including ICT, cloud computing, smartphone apps and advanced data analysis techniques. Patient data can be shared in real time across different departments of a smart hospital or even across smart hospitals in different cities or in the same city. Thus, real-time decisions can be made about a patient's health status and appropriate medications.

Source: Giffinger (2015), Ryba (2017), Mohanty et al. (2016).

According to the holistic definition given by the International Organization for Standardization (ISO), a Smart City is a city that enables economic, social and environmental outcomes, while responding to contemporary challenges (such as rapid population growth, climate change, political and economic instability) by primarily improving the way all areas of urban operations and systems function, engaging the public, using information (and data) through modern technology, and applying collaborative leadership methods to deliver better services and improve the quality of life for citizens (including residents, visitors and businesses)

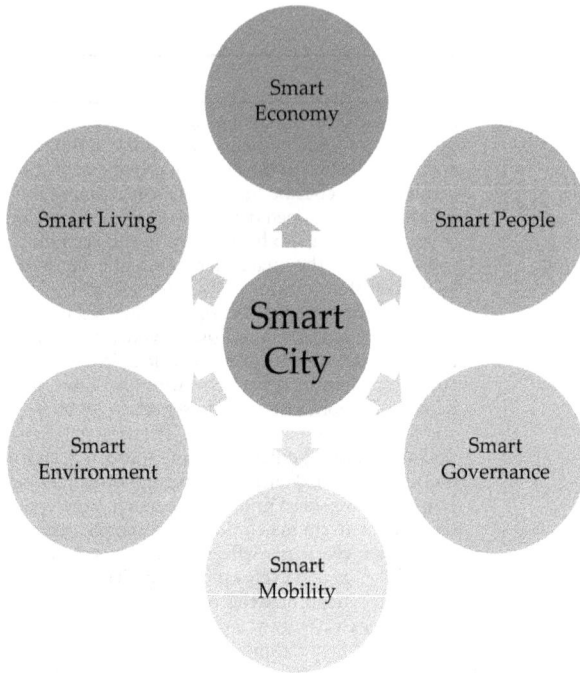

Figure 1.1 Main components of the Smart City concept.

Source: Own study.

without unfairly disadvantaging others or degrading the environment, planning for both the present times and the foreseeable future (Samarak-kody et al., 2022).

Nam and Pardo (2011) explored possible meanings of the term 'smart' in the context of a Smart City. In particular, in marketing language, 'smartness' is a friendlier term than the more elitist term 'intelligent', which is limited to having a quick mind and responding to feedback. Other interpretations suggest that the word 'smart' includes the term 'intelligent', since the idea of smartness is only realized when a smart system adapts to users' needs. Harrison et al. (2010), in an IBM corporate paper, stated that the term 'Smart City' means "an instrumented, connected and intelligent city" (Harrison et al., 2010).

Thus, cities can be defined as 'smart' when they have human and social capital, traditional and modern communication infrastructure (transportation and communication technologies, respectively). Their development is in line with the theory of sustainable development, and participatory governance ensures a better quality of life. It is worth noting that the reference mainly to modern technologies will not be the only factor that binds all these 'smart' factors together. Other very important factors influencing whether a city can be called smart are hard-to-grasp factors such as creativity, innovation and democracy (Qayyum et al., 2021).

In urban planning, the term 'Smart City' is strongly associated with strategic directions for city development. Governments and public agencies at all levels are adopting the term smartness to distinguish their policies and programs aimed at sustainable development, economic growth, a better quality of life for their citizens and improving the well-being and happiness of their residents (Ballas, 2013). Table 1.2 summarizes some sample definitions of Smart City used in the described area.

Table 1.2 Selected Smart City definitions

Author	Year	Definition
Hall	2000	A city that monitors and integrates conditions of all of its critical infrastructures, including roads, bridges, tunnels, rails, subways, airports, seaports, communications, water, power and even major buildings, can better optimize its resources, plan its preventive maintenance activities and monitor security aspects, while maximizing services to its citizens.
Giffinger et al.	2007	A city performing well in a forward-looking way in economy, people, governance, mobility, environment and living, built on the smart combination of endowments and activities of self-decisive, independent and aware citizens. Smart City generally refers to the search and identification of intelligent solutions, which allow modern cities to enhance the quality of the services provided to citizens.
Chen	2010	Smart Cities will take advantage of communications and sensor capabilities sewn into the cities' infrastructures to optimize electrical, transportation and other logistical operations supporting daily life, thereby improving the quality of life for everyone.
Gartner	2011	A Smart City is based on intelligent exchange of information that flows between its many different subsystems. This flow of information is analyzed and translated into citizen and commercial services. The city will act on this information flow to make its wider ecosystem more resource-efficient and sustainable. The information exchange is based on a smart governance operating framework designed to make cities sustainable.
Kominos	2011	Smart Cities as territories with high capacity for learning and innovation, which are built in the creativity of their population, their institutions of knowledge creation and their digital infrastructure for communication and knowledge management.

(Continued)

Table 1.2 (Continued)

Author	Year	Definition
Nam and Prado	2011	Smart City infuses information into its physical infrastructure to improve conveniences, facilitate mobility, add efficiencies, conserve energy, improve the quality of air and water, identify problems and fix them quickly, recover rapidly from disasters, collect data to make better decisions, deploy resources effectively and share data to enable collaboration across entities and domains.
Barrionuevo et al.	2012	Smart City as a high-tech intensive and advanced city that connects people, information and city elements using new technologies in order to create a sustainable, greener city, competitive and innovative commerce, and an increased life quality.
Kourtit et al.	2012	Smart Cities have high productivity, as they have a relatively high share of highly educated people, knowledge-intensive jobs, output-oriented planning systems, creative activities and sustainability-oriented initiatives.
Marsal-Llacuna et al.	2014	Smart Cities initiatives try to improve urban performance by using data, information and information technologies (IT) to provide more efficient services to citizens, to monitor and optimize existing infrastructure, to increase collaboration among different economic actors and to encourage innovative business models in both the private and public sectors.

Source: Albino et al. (2015), Barrionuevo et al. (2012), Chen (2010), Giffinger et al. (2007), Hall (2000), Kominos (2011), Kourti et al. (2012), Nam and Prado (2011), Marsal-Llacuna et al. (2014).

The current Smart City concept is referred to as 3.0. Table 1.3 summarizes the most important stages of its evolution from 1.0 to 3.0. Smart City 1.0 is the first stage of Smart City development, and in this case, the giants of the IT and telecommunications services sector, who sold their off-the-shelf solutions, were most often responsible for SC development. At this stage of development, little attention was paid to whether the creation of Smart City solutions realistically solves any problems for residents. The stage was based on fascination with the latest technologies and the artificial stimulation of demand for them (Shen et al., 2018).

In the case of Smart City 2.0, local authorities become the initiators of change. They look for appropriate solutions to solve existing problems, which is expected to translate directly into improved quality of life for residents. The selection of appropriate technologies and partners is carried out in a much more deliberate manner, and in general, this generation of Smart City is now dominant. Their hallmark remains a large number of projects and programs related to the application of diverse technological solutions, ranging from the promotion of green propulsion sources to systems based on Big Data and publicly accessible Wi-Fi networks.

Smart City 3.0 relies on encouraging residents to use all available technologies, with a huge impact on their attitudes and behaviors through ongoing educational campaigns and promoting the co-creation of Smart City solutions. Undoubtedly, such activities bring numerous benefits for everyone and are a way to effectively improve the quality of life in cities.

A key role in the context of the development of cities of the future is also played by business, i.e., companies involved in the introduction of smart solutions in public spaces, homes or people's lives. These can be small companies (mainly startups), using mainly technology, creativity and access to open data, but also human needs to create software to improve the quality of life, increase productivity and bridge the gap between the city and traditional business in creating social innovation. Another type of companies are those that produce hardware and software

Table 1.3 From Smart Cities 1.0 to Smart Cities 3.0

Smart City concept	Characteristic
Smart Cities 1.0: Technology driven	Smart Cities 1.0 are characterized by technology providers encouraging the adoption of their solutions in cities that have not really been prepared to properly understand the implications of technological solutions or how they can affect the quality of life of citizens. Smart Cities 1.0 are also the philosophy behind most Smart City projects proposed around the world, from PlanIT in Portugal to Songdo in South Korea. These visions of cities of the future are being driven by private sector technology companies, such as Living PlanIT and Cisco. In his book, *Smart Cities*, Anthony Townsend (2015) provides a thoughtful critique of Smart Cities 1.0, arguing that the technology-driven futuristic urban vision misses a key dynamic – how cities interact with their citizens.
Smart Cities 2.0: Technology enabled, city-led	This phase was a response to the over-dominance of technology providers. In this generation, city governments – led by mayors and city administrators – are taking the lead in defining the future of Smart Cities and the role played by the implementation of smart technologies and innovations. In this phase, city administrators are increasingly focusing on technological solutions as enablers to improve the quality of life. During this development period, most of the leading Smart Cities are evolving toward Smart Cities 2.0. Barcelona, for example, is activating more than 20 program areas and more than 100 active Smart City projects from Wi-Fi in public spaces and public transit to smart lighting and promoting electric vehicle charging infrastructure.

(Continued)

Table 1.3 (Continued)

Smart City concept	Characteristic
Smart Cities 3.0: Citizen co-creation	Recently, a new model has begun to emerge. Rather than a technology delivery approach (Smart Cities 1.0) or the use of technology to improve the quality of life of residents (Smart Cities 2.0), leading Smart Cities are beginning to include all urban stakeholders in the creation and implementation of Smart City solutions.
	Vienna, for example, is a leading city that regularly tops the annual Smart City rankings. It continues to be quite active in the 2.0 model. It has a number of Smart City projects launched, but some of these projects have a unique, distinctive character. For example, in a partnership with local energy company, Wien Energy, Vienna has engaged citizens as investors in local solar power plants. This is their individual contribution to the city's goals of developing renewable energy and a zero-carbon economy by 2050.
	Smart City 3.0 also has a strong focus on citizen engagement on affordable housing and gender equality. In this regard, Vancouver conducted one of its most ambitious co-creation initiatives, involving 30,000 citizens in the co-creation of the Vancouver Greenest City 2020 action plan.
	Smart Cities 3.0 are also not just a project for cities operating in developed countries. A very important example in this regard is Medellin, winner of the Urban Land Institute's Innovative City of the Year award. Medellin has focused on the city's grassroots regeneration, engaging citizens from the most distressed neighborhoods in transformative projects, such as a cable car and electric stairs, as well as new high-tech schools and libraries. Medellin has recently expanded its commitment to civic innovation by supporting the development of an impressive innovation district (Ruta N) to attract and retain entrepreneurial talent.
	Smart Cities 3.0 appear to be more embedded in issues of equity and inclusion. The emergence of many sharing initiatives is one manifestation of such an evolution. Many other projects, such as Repair Cafes, libraries that lend tools for home repairs and bike-sharing services, have the potential to not only optimize the use of urban resources but also improve the quality of life for all residents through this process. Cities, such as Amsterdam and Seoul, seem to be leading the way in promoting sharing activities and supporting sharing startups.

Source: Cohen (2015).

to increase energy efficiency, based on smart algorithms (e.g., smart homes), complementing smart grids, or energy solutions to minimize energy losses (so-called smart grid) (Weir, 2015).

The original triple helix analyzes the key role of interaction between the three main helixes of the innovation system: university, industry and government (Wolniak and Jonek-Kowalska, 2022). The quadruple helix model represents a new model of social dynamics based on networking, breaking down barriers between institutions and the integration/collaboration of different social sectors (Klasnic, 2016). The innovation literature sees it as a systemic, open and user-centered model of knowledge creation between government, industry, universities and citizens (Arnkil et al., 2010).

The Smart City 3.0 concept promotes the citizen as an active partner in providing ideas and innovation (March and Ribera-Fumaz, 2016; de Waal and Dignum, 2017). Innovation activities are largely decentralized, moving from a centralized, privileged and closed triple helix of private, government and academic experts to include nontraditional players from among citizens (Capdevila and Zarlenga, 2015). Citizens can play many roles in a Smart City. These include providing feedback on project proposals, directly proposing visions and ideas, participating in decision-making and playing an empowering role as co-creators (Cardullo and Kitchin, 2019). Because the quadruple helix model puts all partners on an equal footing, its proponents assert that it can provide a guiding structure for Smart City governance. Moreover, as stated by Borkowska and Osborne (2018), the quadruple helix exposes citizens as a key factor in evaluating technological innovations and the benefits of Smart City activities, since only they, as the main stakeholders, can undertake this task (Paskaleva et al., 2021).

Government datasets, according to the quadruple helix model, can support citizen engagement and participation in Smart City (Gooch et al., 2015), as they can help identify problems more effectively (Baccarne et al., 2014; Kitchin, 2015). The implementation of the quadruple helix model and its potential success in Smart City projects depend on the ability and willingness of stakeholders to take a role in shaping and pursuing shared benefits. A growing body of literature suggests that universities play a central role in initiating and monitoring urban sustainability initiatives (Karvonen et al., 2018).

The quintuple helix model adds a fifth dimension to innovation processes by emphasizing natural and social ecology (Carayannis and Campbell, 2010). A question that typically remains unresolved within this model, however, is how to connect the five helixes in the innovation process. Markard et al. (2012) argue that this link should be ecological issues, as they relate to social and environmental concerns. These issues are the drivers for future knowledge and innovation (Tratori et al., 2021).

Multinational technology corporations with the ambition to implement the solutions necessary for the cities of the future are playing an increasingly important role in the development of Smart Cities. Many of them have their own Smart Cities development strategies and definitions, such as the Hitachi Corporation, a Japanese technology giant and one of the world's leaders in supporting Smart City solutions. Hitachi defines

Smart Cities as entities that are environmentally conscious, use technological solutions to use energy efficiently and care about the standard of living of their residents, i.e., ensuring a balance between human quality of life and responsibility for aspects of nature (Yoshikawa, 2012).

Areas where smart solutions are particularly common in cities include (Fitsilis, 2022; Lim et al., 2022; Rodrigues and Bubri, 2022):

- public transportation;
- traffic management;
- public offices and services;
- energy;
- urban infrastructure;
- community development;
- health.

The above catalog is also complemented by specific solutions, such as (Fitsilis, 2022; Lim et al., 2022; Rodrigues and Bubri, 2022;):

- parking spaces (you can use the app to search for free parking spaces in the city);
- use of renewable energy sources;
- energy self-sufficient buildings;
- apps that allow users to buy tickets, pay for parking spaces and use services at government offices;
- measuring water and energy consumption;
- garbage disposal;
- Smart City lighting;
- e-payments;
- citizen budgets;
- bicycles and city scooters.

Table 1.4 summarizes examples of real-world Smart Cities solutions.

Smart transportation is very often written about in the context of Smart City. This is not without reason, as traffic jams, crowding, and noise and pollution are problems for many modern cities. The introduction of comprehensive solutions in this area has allowed many authorities to deal with these problems.

The effects of the use of intelligent transportation, as assumed, are the following (Fitsilis, 2022; Lim et al., 2022; Rodrigues and Bubri, 2022;):

- improving the flow of urban traffic;
- increasing the comfort of movement;
- reducing the stress of urban traffic;
- greening the various forms of transportation.

Table 1.4 Examples of Smart Cities activities

City	Description of initiatives
Vienna	Replacing traditional buses with electric ones that use energy mainly from renewable sources. Creating an app for passengers to keep track of traffic.
Copenhagen	Integrated pedestrian, bicycle and car transportation (Park&Ride, Bike&Ride). Closure of the city center to automobile traffic.
London	Organizing vehicular traffic by introducing a toll system for moving around the city center. Introducing the SCOOT traffic management system, which controls the operation of traffic lights at intersections.
Madrid	The creation of the MiNT Madrid iNTeligentne/Smarter Madrid platform to facilitate the management of city services, such as garbage collection, recycling and green space organization.
Luxembourg	The e-City initiative that had two goals. The first was to make high-quality city spaces as accessible as possible to residents. The second was to ensure the integration of the city's community and enable its further development. The first mobile payment services for citizens were based on SMS technology (e.g., μ-Payments, SMS4Ticket, Call2Park), and key contracts were executed by all national telecommunications operators to maintain the neutrality of public services.

Source: Sowa (2022), Nathansohn and Lahat (2022).

One of the standard examples of smart urban transportation is transportation based on streetcars, subways, as well as hybrid and electric buses. This reduces emissions – especially in the inner city. An integrated traffic system takes into account the needs of all traffic participants – pedestrians, cyclists and motorists. In a Smart City, passengers and drivers find the fastest way to get to their desired destination thanks to apps. Meanwhile, the management system continuously analyzes the city's traffic and adjusts light cycles according to its intensity.

1.2 Sustainability as a priority for Smart City development

The rapid growth of urban populations and the associated increase in resource consumption will inevitably create numerous challenges for modern cities. This fact is emphasized by paradigm shifts in how cities operate, including, in particular, sustainability. Allen and Hoekstra (1993) emphasize the importance of establishing the scale, at which the sustainability of urban systems is assessed. Achieving sustainability at the global scale requires totally different types of actions than at the urban scale (Bibri and Krogstie, 2017). However, there is no concretized and universal definition of sustainability at the urban scale, but a set of characteristics of urban sustainability can be identified (Barrionuevo et al., 2012). These include intergenerational equity, intra-generational equity (social, geographic and governance equity), environmental protection, significant reduction in the use of nonrenewable resources, economic vitality and diversity, community autonomy, citizen well-being and satisfaction of basic human needs (Maclaren, 1996).

These features include three dimensions of sustainability: environmental, economic and social (Lehtonen, 2004). The environmental dimension refers to the ecological aspect and includes the protection of the environment (flora and fauna) and natural resources, as well as an economy based on energy production. The social dimension includes justice, community autonomy, the well-being of citizens and the satisfaction of basic human needs. The economic dimension, on the contrary, consists of economic vitality and urban diversity (Batty et al., 2012; Tripathi et al., 2022).

An interesting review of Smart City definitions in the context of sustainability was made by Toli and Mrtagh (2020). According to their research, environmentally oriented definitions tend to focus on the impact that digital technologies will have on individual city services. Such technologies can be used to improve resource use and reduce emissions (O'Grady and O'Hare, 2012). This can lead not only to the development of smarter transportation infrastructure, improved water supply and waste disposal systems, as well as more efficient thermoregulation of buildings, but also to improved city administration services, safer public spaces and a better response to the needs of an aging population (European Commission, 2019).

Numerous industry players, operating mainly in the IT sector, have provided definitions similar to the one above, developed by the European Commission. Microsoft (2018) considered a Smart City to be a place, where ICT is used to improve the delivery of services to citizens, such as the provision of energy and water, public safety and transportation, the delivery of health services, or taking care of the health, sustainability, resilience and security of cities.

Bosch Global (2019) provided a similar definition, exposing that the use of various technologies can improve citizens' overall quality of life by saving time, using new methods of mobility and breathing cleaner air, and leading to a reduction in traffic, as well as the development of smart homes and energy-efficient use of buildings (Viitanen and Kingston, 2014). Nevertheless, citizens' quality of life and more comfortable, safe and convenient lifestyles should be in harmony with the environment, and the goal of Smart Cities should be to enable a well-balanced relationship between people and the Earth (Hitachi, 2012; Aoun, 2013).

All sustainability-oriented Smart City definitions also include a social dimension (Caragliu et al., 2011). When the Smart City concept was introduced, it was viewed as a strategic tool to emphasize the growing importance of ICT, as well as social and environmental capital in shaping the competitiveness of modern cities (Schaffers et al., 2012). As a result, Smart City definitions that include the environmental dimension of sustainability often also include the social dimension directly or indirectly. Schaffers et al. (2012) argue that this is due to the distinctive attributes that social and environmental capital can offer to Smart Cities as opposed to their 'more technological counterparts', referred to in the literature as digital or Smart Cities. According to the above, the main difference between digital Smart Cities and sustainable Smart Cities stems from the predominance of the human-social element in defining and operating the latter.

Sustainable urban development is primarily about shaping the appropriate relationships taking place between urban systems (economic, social, natural systems) and the relationships of these systems with the environment. The causal role in shaping these relationships is played by people, represented both in the city government and other forms of urban stakeholders that make up the local community (McFarlane and Söderström, 2017).

The topic of sustainable urban development is also addressed in the 'Sustainable Development Goals 2030' – a document published by the United Nations (UN). The goals contained in the study were adopted by each UN Member State in 2015 and should be met by 2030. However, they are general and universal in nature and, like the set of ISO standards, are not legal standards. The UN study aims to "make cities and human settlements safe, stable, sustainable and inclusive". Goal 11 is, therefore, closely related to sustainable development and raising the standard of living in cities, which is at the heart of the Smart City concept. It includes the following specific issues (Korenik, 2019):

- ensuring access to adequate, safe and affordable housing and basic services;
- improving living conditions in slums;
- ensuring access to safe transportation systems (with a concomitant increase in road safety);
- ensuring sustainable urbanization and participation in integrated and sustainable planning and management of human settlements;
- strengthening the protection and safeguarding of the world's cultural and natural heritage;
- reducing the city's per capita negative environmental impact rate;
- ensuring easy and safe access to green spaces;
- increasing the number of cities benefiting from the development and implementation of integrated policies and plans seeking to increase inclusiveness and resource efficiency.

Measures to stimulate and maintain sustainable urban development should mainly consist of development of sustainable transportation; revitalization; recycling; low greenhouse gas emissions, ultimately aimed at zero carbon; development of green areas; formation of environmentally friendly urban governance; and efficient energy management.

The concept of smart environment strictly refers to the idea of sustainable development and is related to environmental techniques and systems, such as optimization of energy consumption, use of renewable energy sources instead of non-renewable ones, low-carbon policies, environmental education or a system for reducing water consumption. Resource-efficient management primarily requires stimulating eco-innovation, i.e., forms of innovation that result in (or aim at) significant and visible progress toward sustainable development. The general principles, inherent in the area of smart environment, are the greening of the economy (taking into account the requirements of environmental protection in all areas of economic activity), avoidance of pollution (prevention), economic efficiency (achieving environmental goals at minimal social cost), as well as openness of information as a basis for building cooperation and decision-making between participants in the development process.

The Smart City should strive for energy self-sufficiency. The public, as a group of stakeholders in the city's environmental policy, must be environmentally aware and understand the rationale for environmental policy and actively participate in it. Economic entities, often themselves perpetrators of environmental pollution, are another important stakeholder of the policy. If the introduced laws are transparent and restrictive, these entities will take action to protect the environment, often generating additional competitive advantages in the process.

An example of combining the concepts of sustainability and Smart City can be seen in the actions taken by the City of Amsterdam. Amsterdam's Smart City paradigm involves a shift from centralization to decentralization, from a top-down approach to a bottom-up one, and the creation of assumptions for activities based on data flow. It envisions a shift from a linear economy (a type of economy, in which there is production, use and discard, resulting in the generation of waste on the one hand and the depletion of natural resources on the other hand) to a circular economy (closed loop, CE), which is capable of reproducing the life cycle of production and consumption of waste, which can then be used as primary raw materials or in the production process. It implements the 3R principle, i.e., reduce-recycle-reuse. This means that any waste can become raw materials at the same time, which changes the management paradigm. Table 1.5 shows examples of urban Smart City projects implemented by Amsterdam that directly revolve around sustainability.

Table 1.5 Selected Smart City projects in Amsterdam relating to sustainability

Smart City area	Project with its characteristic
Digitization	STEORA Smart City benches – with Internet access, a charging station for phones; operating thanks to solar energy; containing built-in sensors that acquire information on, for example, weather conditions and the number of users of its components; built with materials that reduce vandalism.
Energy	SmartCrusher – a technology that allows recycled concrete to be reused, such as for building construction, reducing CO_2 emissions by about 70%.
Mobility	Foodlogica – a logistics service, with which to deliver food to a selected point in the city in an environmentally friendly, CO_2-emission free, using eco-friendly tricycles.
Closed-loop economy	Solar-powered smart trash cans – automatically compress waste, so they can hold eight times more than traditional trash cans; report on fill status; some have sensors that detect certain types of substances.
Residents and life	TreeWiFi – birdhouses with built-in LED lights indicating the current level of air pollution, if it is low the birdhouses provide free Wi-Fi.

Source: Amsterdam Smart City (2022), Korenik (2019), Somayya and Ramaswamy (2016), Mora and Bolici (2015), Danielou (2014).

1.3 Evaluation and contemporary challenges of a Smart City

Various approaches to measuring the involvement of cities in the implementation of the Smart City concept can be found in the literature. The system of such assessment can be based on a three-stage structure referring to the considerations presented in the previous subsections of this monograph. Its proposed structure is shown in Figure 1.2 and Table 1.6.

Issues related to the benefits of evaluating the involvement of cities in the implementation of the Smart Cities concept were described by Patrão et al. (2020). The most important potential benefits of evaluating the implementation of the Smart City concept for individual stakeholders are presented in Table 1.7.

According to Sharifi (2019), there are two main approaches to studying Smart City assessment tools: one, focusing on providing an overview that includes descriptive characteristics, and another, focusing on providing an overview that includes more detailed thematic and indicator analyses. Albino et al. (2015) presented an overview of Smart City assessment systems up to 2015. Their analysis is quantitative regarding the indicators

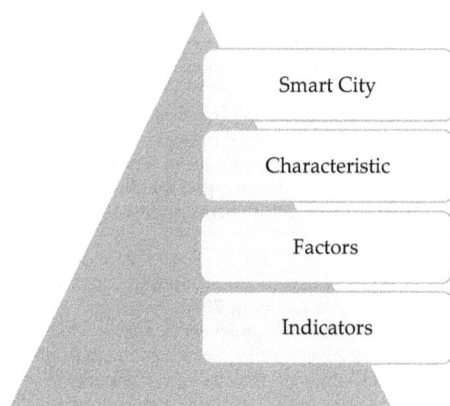

Figure 1.2 Operationalizing Smart City concept – structure.
Source: Giffinger et al. (2007).

Table 1.6 Examples of factors to measure particular dimension of a Smart City

Category	*Characteristic*
Smart economy	• Innovative spirit • Entrepreneurship • Economic image and trademarks • Productivity • Flexibility of labor market • International embeddedness • Ability to transform
Smart mobility	• Local accessibility • (Inter-)national accessibility • Availability of ICT infrastructure • Sustainable, innovative and safe transport systems
Smart environment	• Attractiveness of natural conditions • Pollution • Environmental protection • Sustainable resource management
Smart people	• Level of qualification • Affinity to lifelong learning • Social and ethnic plurality • Flexibility • Creativity • Cosmopolitanism/open-mindedness • Participation in public life
Smart living	• Cultural facilities • Health conditions • Individual safety • Housing quality • Education facilities • Tourism attractiveness • Social cohesion

(*Continued*)

Table 1.6 (Continued)

Category	Characteristic
Smart governance	• Participation in decision-making • Public and social services • Transparent governance • Political strategies and perspectives

Source: Giffinger et al. (2007).

Table 1.7 Evaluating the benefits of implementation of the Smart City concept

Stakeholder	Potential benefits
Cities and municipal authorities	• Monitor performance to improve the international image and competitive position of the city in the eyes of investors, as well as to improve the situation of creative residents and the city's position in the eyes of investors, residents and societies. • Justify the value of Smart City investments. • Identify strengths and weaknesses and guide Smart City planning. • Track progress toward predefined goals and objectives and determine the city's position in its smartness efforts. • Understand the socioeconomic and environmental implications of Smart City projects. • Understand the technical requirements of Smart City projects. • Learn from the experiences of peers (when assessment includes benchmarking). • Identify and show cases of best practices, from which lessons can be learned. • Increase management transparency. • Stimulate discussions among various stakeholders that can result in better resource allocation.
Investors and financial agencies	• Evidence-based evaluation of completed or ongoing projects. • Scientific ways to prioritize the allocation of funds. • Increased ability to make decisions about the best sites for future investments. • Ability to identify and capitalize on new business opportunities.
Researchers	• Develop new strategies to improve Smart City performance. • Simplify the complexity of the Smart City concept.
Inhabitants	• Increased awareness of the benefits of Smart City projects. • Ability to make informed decisions when it comes to future investments. • Motivation to engage in smart city development activities and communicate their desires and priorities to city authorities.

Source: Patrão et al. (2020), Giffinger (2010), Caird et al. (2016), Garau and Pavan (2018), Fernandez-Anez et al. (2018), Sang et al. (2015), Debnath et al. (2014).

used with little qualitative detail, so it is not possible to assess the potential gaps of the assessment tools presented. A more detailed summary of the various methods used for Smart City assessment can be found and used in the publication by Patrão et al. (2020).

The tools for assessing Smart City development involve various indicators. Nevertheless, according to the authors of this chapter, developers of SC assessment methodologies should take into account that both the proposed tools and the results obtained using them should be easy to use and understand by urban stakeholders, including, most importantly, city managers. Huovila et al. (2019) also stated that further development of indicators and evaluation principles must be grounded in an analysis of urban needs. This is a pertinent and important point, as sometimes the indicators used in evaluating cities do not correspond with the directions of their development.

The CIMI (Cities in Motion Index) is one of the most important when it comes to Smart Cities. It is used by Profs. Joan Enric Ricart and Pascual Berrone of the IESE Center for Globalization and Strategy at the Spanish Business School. In the CIMI methodology, Smart City assessment is carried out by taking into account the following factors (IESE, 2020):

- human capital that illustrates what prospects the city provides for people and whether it supports their development and stimulates creativity (this factor is determined, among other things, by the share of residents with secondary or higher education, the number of cultural buildings in the city or the number of universities with top educational rankings);
- social cohesion relating to crime, the quality of residents' health, their level of satisfaction with life or issues of exclusion based on age, gender, etc.;
- economics defining the conditions for business development, fostering innovation and entrepreneurship, as well as defining the well-being of residents in terms of wages, the purchasing power of money or the level of economic development;
- management relating directly to the efficiency, quality and capacity of the local government, the city's finances, possession of ISO 37120 certification, research facilities, open access platforms or corruption index;
- environment including assessment of the greening of housing, use of alternative energy sources, waste management, water and wastewater management, as well as air quality and pollution of the ecosphere;
- mobility and transportation regarding the quality of infrastructure and whether residents are able to get around the city comfortably and quickly (specific indicators: the number of services that allow bicycle rentals, the city's congestion rate or the number of vehicles per household);
- spatial planning that takes into account how the city is urbanized and adapted to the needs of the population;

- international reach concerning the city's international recognition (although the number of McDonald's restaurants, for example, is also taken into account here);
- technology relating to the level of technological expertise among residents, the level of technology development in the city and access to the latest Smart City solutions.

The last available full version of the assessment is the 2020 version. Examples of indicators for each area from the aforementioned report are included in Table 1.8.

Table 1.8 CIMI assessment indicators

Factors	Indicators
Human capital	Secondary or higher education
	Schools
	Business schools
	Expenditure on education
	Per capita expenditure on leisure and recreation
	Movement of students
	Museum and art galleries
	Number of universities
	Theaters
Social cohesion	Female-friendly
	Hospitals
	Crime rate
	Slavery index
	Happiness index
	Gini index
	Peace index
	Health index
	Price of property
	Homicide rate
	Death rate
	Female employment ratio
	Suicide rates
	Unemployment rate
	Terrorism
Economic	Collaborative economy
	Ease of starting a business
	Mortgage
	Motivation that people have to undertake early-stage entrepreneurial activity
	Number of headquarters
	Purchasing power
	Productivity
	Hourly wage in US dollars
	Time required to start a business
	GDP
	GDP per capita
	Estimated GDP

(Continued)

Table 1.8 (Continued)

Factors	Indicators
Governance	Government building
	E-Government Development Index (EGDI)
	Embassies
	Employment in the public administration
	Strength of legal rights index
	Corruption perceptions index
	ISO 37120 certification
	Research centers
	Open data platform
	Reserves
	Reserves per capita
Environmental	Solid waste
	Future climate
	CO_2 emissions
	Methane emissions
	Environmental performance index
	CO_2 emission index
	Pollution index
	PM_{10}
	$PM_{2.5}$
	Percentage of the population with access to the water supply
	Renewable water resources
Mobility and transportation	Bicycle rental
	Moped rental
	Scooter rental
	Bicycles per household
	Bike sharing
	Traffic inefficiency index
	Exponential traffic index
	Traffic index
	Length of the metro system
	Metro stations
	High-speed train
	Commercial vehicles in the city
	Flights
Urban planning	Bicycles for rent
	Buildings
	Number of people per household
	Percentage of the urban population with adequate sanitation services
	Buildings over 35 meters high
International projection	Number of passengers per airport
	Hotels
	Restaurant index
	McDonald's
	Number of conferences and meetings
	Number of photos of the city uploaded online

(*Continued*)

Table 1.8 (Continued)

Factors	Indicators
Technology	3G coverage
	Innovation index
	Internet
	Online banking
	Online video calls
	LTE/WiMAX
	Mobile phone penetration ratio
	Personal computers
	Social networks
	Landline subscriptions
	Broadband subscriptions
	Telephony
	Mobile telephony
	Internet usage away from home and/or office
	Internet speed
	Web Index
	Wi-Fi hotspots

Source: IESE (2020).

In the latest version of the 2020 ranking, the ten cities in the world, most committed to implementing Smart Cities, are (IESE, 2020):

• London (CIMI 100);
• New York City (CIMI 94.63);
• Amsterdam (CIMI 86.7);
• Paris (CIMI 86.23);
• Reykjavik (CIMI 85.35);
• Tokyo (CIMI 84.11);
• Singapore (CIMI 82.73);
• Copenhagen (CIMI 81.8);
• Berlin (CIMI 80.88);
• Vienna (CIMI 78.85).

Table 1.9 provides brief characteristics of the achievements of the listed cities in implementing the Smart City concept.

Many cities today have ambitions to become Smart Cities, as evidenced by the popularity of various rankings. To achieve this, however, they must overcome the challenges of developing a complex strategy that includes public and private participants, direct and indirect stakeholders, integrators, network and managed service providers, product vendors and IT infrastructure providers.

Table 1.9 Characteristics of the top ten cities included in IESE (2020)

City	Characteristic
London	London is a great place for business development, which is reflected in the significant number of startups, developers and researchers, who come and work in the city. The open data platform, London Datastore, and investments in expanding a well-organized transportation network are also significant.
New York City	New York City remains a leader when it comes to economic development, as expressed in GDP levels. The city is also home to nearly 7,000 high-tech companies. Add to this the presence of an open Wi-Fi network, and you get a picture of a city that relies primarily on technological development.
Amsterdam	A unique example of a green city, with 90% of households using bicycles. It is also impossible to forget the impressive number of modern and leading FinTech companies. Amsterdam is also currently projected to be the first zero-emission city in Europe.
Paris	The French capital is home to nearly half of France's companies. Local authorities are encouraging the use of ride-sharing services, emphasizing their concern for the environment. Add to that the Grand Paris Express project, which will add four more subway lines, 68 new train stations and 200 km of rail lines.
Reykjavik	In the case of Reykjavik, its high position is influenced by the care for the environment. More than 90% of its electricity is generated by hydroelectric and geothermal power plants. The Icelandic capital has also committed to becoming a zero-emission city by 2040.
Tokyo	The capital of the Land of the Rising Sun is the most innovative city in the world and has also remained in the top ten of the Global Financial Centers Index for years.
Singapore	The city provides the fastest Internet and almost all residents use cell phones, hospitals are already staffed with robots and autonomous cabs can be found on the streets. Undoubtedly, there is a reason why Singapore remains the best-rated city in terms of available technology.
Copenhagen	The city has a balanced development strategy that is paying off. The city has set out to achieve zero-carbon emissions as early as 2025.
Berlin	The German capital is committed to sustainability in all areas and ranks in the top 50 in every category, noting, however, the worst score in terms of the economy.
Vienna	The Austrian capital was ranked among the world's top ten Smart Cities for the first time in 2019, and in 2020, the city repeated the result. In this case, the city stands out for its very strong position in the mobility and transportation category.

Source: IESE (2020), Lewandowski (2021).

Cities wishing to engage in the Smart City concept have to face numerous challenges, which can be addressed today through a combination of technological innovation and cooperation between public organizations and private companies. Among the most important of these are (Beevor, 2018; Stone, 2018; Joshi, 2019):

- Infrastructure – Smart Cities use sensor technology to collect and analyze information to improve the quality of life for residents. Sensors collect data on everything, from rush hour traffic statistics, to crime rates, to air quality assessments. Installing and maintaining these sensors requires complex and expensive infrastructure. Large metropolitan areas already face the challenge of replacing decades-old infrastructure, such as underground wiring, steam pipes and transport tunnels, and installing high-speed Internet. Broadband wireless services are becoming more common, but there are still areas in large cities where access is limited.
- Security and hackers – With the development of IoT technology and the number of sensors in use, digital security threats are increasing. This begs the question: is technology really considered 'smart' if hackers can break into it and paralyze an entire city? Smart Cities are investing more and more money and resources in security, while technology companies are creating solutions with new mechanisms to protect against hacking and cybercrime.
- Privacy concerns – In every major city, there is a balance between quality of life and reduced privacy. While everyone wants to enjoy a more comfortable, peaceful and healthier environment, no one wants to feel like they are constantly being monitored. Cameras installed on every street corner can help deter crime, but they can also create fear and paranoia. Another major problem is the amount of data collected from all the smart sensors that residents encounter every day. When creating Smart City solutions, it is important to think about the community and consider how it might respond to new technology.
- Educating and engaging the community – For a Smart City to truly exist and thrive, it needs citizens who are engaged and actively using new technologies. For any new urban technology project, part of the implementation process must be educating the community about its benefits. This can be done through a series of in-person meetings at city hall, e-mail campaigns, and an online education platform to keep residents informed.
- Being socially inclusive – It is important that Smart City planning should take into account all groups of people, not just the wealthy and technologically advanced but also people with disabilities and the elderly who have difficulty operating modern devices. Technology

should always work to bring people together, not further divide them based on income or education level. Thinking about these communities will promote the overall success of the solution beyond the realm of tech-savvy users.

• Lack of experienced professionals – Technical experts are needed to prepare strategies to achieve the success of a Smart City project, to identify areas to implement technologies and to operate these tools. Municipal authorities should, therefore, continuously assess and compare the needs in this area with the available resources and take care of the development of human resources for the creation of Smart Cities.

• Inconsistent network connectivity – To intelligently manage the municipality, many sensors, cameras and actuators are being installed everywhere. These sensors collect and transmit large amounts of data in real time. Analysis and processing of the collected data should take place almost instantaneously to enable efficient management of city operations. For instant processing, high-speed Internet connectivity is essential. Currently, 4G cellular coverage systems are available, which are not efficient enough for high-speed data transfer. Therefore, this issue should be obligatorily included in Smart City development strategies.

• Financial constraints – Transforming an ordinary city into a smart one requires a large budget. One of the reasons is the lack of understanding of the financial dimension of the IT projects being implemented. An important issue also relates to the necessary public-private partnerships. Such partnerships are required, but sometimes difficult to obtain and manage, and this can be one of the constraints to the successful development and implementation of Smart City projects. Meanwhile, in practice, in many cases, it turns out that public procurement processes are not designed for the rapid application of innovative solutions or Smart City projects.

Bibliography

Albino, V., Berardi, U., Dangelico, R.M. (2015). Smart cities: Definitions, dimensions, performance, and initiatives. *Journal of Urban Technology, 22*, 3–21. https://doi.org/10.1080/10630732.2014.942092

Allen, T., Hoekstra, T.W. (1993). *Toward a Definition of Sustainability. Sustainable Ecological Systems: Implementing an Ecological Approach to Land Management.* Rocky Mountain Forest and Range Experiment Station: Fort Collins, CO, 98–107.

Amsterdam Smart City. (2022). https://amsterdamsmartcity.com/, [access data: 9.10.2022].

Aoun, C. (2013). *The Smart City Cornerstone: Urban Efficiency.* Schneider-Electric: London.

Arnkil, R., Järvensivu, A., Koski, P., Piirainen, T. (2010). *Exploring quadruple helix outlining user-oriented innovation models.* Työraportteja, 85, Working Papers. Available at: https://trepo.tuni.fi/handle/10024/65758.

Baccarne, B., Merchant, P., Schuurman, D., Colpaert, P. (2014). Urban socio-technical innovations with and by citizens. *Interdisciplinary Studies Journal,* *3*(4), 143–156.

Ballas, D. (2013). What Makes a 'Happy City'? *Cities, 32*(1), S39–S50.

Barrionuevo, J.M., Berrone, P., Ricart, J.E. (2012). *Smart Cities, Sustainable Progress,* IESE Insight 14, 50–57.

Batty, M., Axhausen, K.W., Giannotti, F., Pozdnoukhov, A., Bazzani, A., Wachowicz, M., et al. (2012). Smart cities of the future. *The European Physical Journal Special Topics, 214,* 481–518. https://doi.org/10.1140/epjst/e2012-01703-3

Beevor, M. (2018). *6 Challenges smart cities face and how to overcome them,* https://statetechmagazine.com/article/2018/12/6-challenges-smart-cities-face-and-how-overcome-them, [access data: 9.10.2022].

Bibri, S.E., Krogstie, J. (2017). Smart sustainable cities of the future: An extensive interdisciplinary literature review. *Sustainable Cities and Society, 31,* 183–212. https://doi.org/10.1016/j.scs.2017.02.016

Borkowska, K., Osborne, M. (2018). Locating the fourth helix: Rethinking the role of civil society in developing smart learning cities. *Internal Review of Education, 64*(3), 355–372.

Bosch Global. (2019). *Smart City Concepts–The City of Tomorrow.* Available online at: https://www.bosch.com/stories/smart-city-challenges/pdf, [access data: 9.10.2022].

Caird, S., Hudson, L., Kortuem, G. (2016). *A Tale of Evaluation and Reporting in UK Smart Cities Other Sally Caird with Lorraine Hudson and Gerd Kortuem.* The Open University: Milton Keynes.

Capdevila, I., Zarlenga, M. (2015). Smart city or smart citizens? The Barcelona Case. *Journal of Strategy and Management, 8*(3): 266–282.

Caragliu, A., Del Bo, C., Nijkamp, P. (2011). Smart cities in Europe. *Journal of Urban Technology, 18,* 65–82. https://doi.org/10.1080/10630732.2011.601117

Carayannis, E.G., Campbell, D.F.J. (2010). Triple Helix, Quadruple Helix and Quintuple Helix and how do knowledge, innovation, and environment relate to each other? *The International Journal of Social Ecology and Sustainable Development, 1,* 41–69.

Cardullo, P., Kitchin, R. (2019). Smart urbanism and smart citizenship: The neoliberal logic of 'citizen-focused' smart cities in Europe. *Environment and Planning C: Politics and Space, 37*(5), 813–830.

Chen, T.M. (2010). Smart Grids, Smart Cities Need Better Networks [Editor's Note]. *IEEE Network, 24*(2), 2–3.

Cohen, B. (2015). *The 3 generations of smart cities,* https://www.fastcompany.com/3047795/the-3-generations-of-smart-cities, [access data: 9.10.2022].

Danielou, J. (2014). *Smart city and sustainable city: The case of Amsterdam,* https://www.citego.org/bdf_fiche-document-2429_en.html, [access data: 9.10.2022].

de Waal, M., Dignum, M. (2017). The citizen in the smart city. How the smart city could transform citizenship. *Information Technology, 59*(6), 263–273.

Debnath, A.K., Chin, H.C., Haque, M.M., Yuen, B. (2014). A methodological framework for benchmarking smart transport cities. *Cities, 37,* 47–56.

European Commission. (2019). *Smart Cities.* Available online at: https:// ec.europa.eu/info/eu-regional-and-urban-development/topics/cities-and-urban-development/city-initiatives/smart-cities_en, [access data: 9.10.2022].

Falco, S., Angelidou, M., Addie, J.-P.D.J.E.U. (2019). Studies, R. From the "smart city" to the "smart metropolis"? Building resilience in the urban periphery. *European Urban and Regional Studies, 26,* 205–223.

Fernandez-Anez, V., Fernández-Güell, J.M., Giffinger, R. (2018). Smart City implementation and discourses: An integrated conceptual model. The case of Vienna. *Cities, 78,* 4–16.

Fitsilis, P. (2022). *Building on Smart Cities Skills and Competencies.* Springer: Cham.

Florida, R. (2014). The creative class and economic development. *Economic Development Quarterly, 28*(3), 196–205.

Garau, C., Pavan, V.M. (2018). Evaluating urban quality: Indicators and assessment tools for smart sustainable cities. *Sustainability, 10,* 575.

Giffinger, R. (2010). Smart cities ranking: An effective instrument for the positioning of the cities? *ACE: Architecture, City and Environment, 4*(12), 7–26.

Giffinger, R. (2015). *European Smart City Model (2007–2015).* Vienna University of Technology, http://www.smart-cities.eu [access data: 9.10.2022].

Giffinger, R., Fertner, R., Kramar, Ch., Kalasek, R., Pichler-LilanCentre of Regional Science, Vienna Utović, N., Meijers, E. (2007). *Smart cities – Ranking of European medium-sized cities,* https://www.researchgate.net/publication/261367640_ Smart_cities_-_Ranking_of_European_medium-sized_cities, [access data: 9.10.2022].

Gooch, D., Kortuem, G., Wolff, A., Brown, R. (2015). Reimagining the role of citizens in Smart City projects. In: *AMC International Conference on Pervasive and Ubiquitous Computing,* Osaka, Japan, 23–25 September. ACM: New York, 12–14.

Hall, R.E. (2000). The Vision of a Smart City, Proceedings of the 2nd International Life Extension Technology Workshop, Paris, France.

Harrison, C., Eckman, B., Hamilton, R., Paraszczak, J., Williams, P. (2010). Foundations for Smarter Cities. *IBM Journal of Research and Development, 54*(4), 5512826.

Hitachi. (2012). Hitachi's vision of the smart city. *Hitachi Review, 61,* 111–118.

Huovila, A., Bosch, P., Airaksinen, M. (2019). Comparative analysis of standardized indicators for Smart sustainable cities: What indicators and standards to use and when? *Cities, 89,* 141–153.

IESE Cities in Motion Index 2020. (2020). https://blog.iese.edu/cities-challenges-and-management/2020/10/27/iese-cities-in-motion-index-2020/, [access data: 9.10.2022].

Jonek-Kowalska, I., Wolniak, R. (2021a). Economic opportunities for creating smart cities in Poland. Does wealth matter? *Cities, 114,* 1–6.

Jonek-Kowalska, I., Wolniak, R. (2021b). The influence of local economic conditions on start-ups and local open innovation system, *Journal of Open Innovations: Technology, Market and Complexity, 7*(2), 1–19.

Jonek-Kowalska, I., Wolniak, R. (2022). Sharing economies' initiatives in municipal authorities' perspective: Research evidence from Poland in the context of smart cities' development. *Sustainability, 14*(4), 1–23.

Joshi, N. (2019). *4 Challenges faced by smart cities,* https://www.allerin.com/blog/4-challenges-faced-by-smart-cities, [access data: 9.10.2022].

Karvonen, A., Martin, C., Evans, J. (2018). University campuses as testbeds of smart urban innovation. In: Coletta, C., Evans, L., Heaphy, L., Kitchin, R. (eds) *Creating Smart Cities.* Routledge: London:, 104–117.

Kitchin, R. (2015). Making sense of smart cities: Addressing present shortcomings. *Cambridge Journal of Regions, Economy and Society, 8*(1), 131–136.

Klasnic, J. (2016). Specific barriers for quadruple helix innovation model development – Case of Croatia. In: *ENTRENOVA Conference,* Rovinj, 8–9 September. IRENET: Zagreb, 399–407.

Komninos, N. (2011). Intelligent cities: Variable geometries of spatial intelligence. *Intelligent Buildings International, 3*(3), 172–188.

Korenik, A. (2019). *Rozwój zrównoważony na przykładzie miast inteligentnych,* https://www.researchgate.net/publication/337608287_Rozwoj_zrownowazony_na_przykladzie_miast_inteligentnych_smart_cities?channel=doi&linkId=5de03c3c299bf10bc32ec9dd&showFulltext=true; https://doi.org/10.13140/RG.2.2.12024.80649, [access data: 9.10.2022].

Kourtit, K., Nijkamp, P. (2012). Smart cities in the innovation age. *Innovation: The European Journal of Social Science Research, 25*(2), 93–95.

Lehtonen, M. (2004). The environmental-social interface of sustainable development: Capabilities, social capital, institutions. *Ecological Economics, 49,* 199–214. https://doi.org/10.1016/j.ecolecon.2004.03.019

Lewandowski, A. (2021). *Top 10 smart cities. Najbardziej inteligentne miasta świata – ranking,* https://almine.pl/najbardziej-inteligentne-miasta-swiata-top-10-ranking/, [access data: 9.10.2022].

Lim, T.C., Wilson, B., Grohs, J.R., Pingel, T.J. (2022). Community-engaged heat resilience planning: Lessons from a youth smart city STEM program. *Landscape and Urban Planning, 226,* 104497.

Lombardi, P., Giordano, S., Farouh, H., Yousef, W. (2012). Modelling the smart city performance. *Innovation: The European Journal of Social Science Research, 25,* 137–149. https://doi.org/10.1080/13511610.2012.660325.

Maclaren, V.W. (1996). Urban sustainability reporting. *Journal of the American Planning Association, 62,* 184–202. https://doi.org/10.1080/01944369608975684

March, H., Ribera-Fumaz, R. (2016). Smart contradictions: The politics of making Barcelona a Self-sufficient city. *European Urban and Regional Studies, 23*(4), 816–830.

Markard, J., Raven, R., Truffer, B. (2012). Sustainability transitions: An emerging field of research and its prospects. *Research Policy, 41,* 955–967.

Marsal-Llacuna, M.L., Colomer-Llina, J., Melendez-Frigola, J. (2014). Lessons in urban monitoring taken from sustainable and livable cities to better address the Smart Cities initiative. *Technological Forecasting and Social Change, 90,* 611–622.

McFarlane, C., Söderström, O. (2017). On alternative smart cities: From a technology-intensive to a knowledge-intensive smart urbanism. *City, 21,* 312–328. https://doi.org/10.1080/13604813.2017.1327166

Merli, M.Z., Bonollo, E. (2014). *Performance Measurement in the Smart Cities.* Springer: Berlin/Heidelberg, 139–155.

Microsoft. (2018). *Powering smart cities with AI and IoT.* Available online at: https://info.microsoft.com/rs/157-GQE-382/images/EN-CNTNT-Info-graphic-PoweringSmartCitieswithAIandIOT.pdf, [access data: 9.10.2022].

Mohanty, S.P., Choppali, U., Kougianos, E. (2016). Everything you wanted to know about smart cities. *IEEE Consumer Electronics Magazine, 5*(3), 60–70.

Mora, L., Bolici, R. (2015). How to become a smart city: Learning from Amsterdam. In: Bisello, A., Vettorato, D., Stephens, R., Elisei, P. (eds) *Smart and Sustainable Planning for Cities and Regions. SSPCR 2015. Green Energy and Technology.* Springer: Cham. https://doi.org/10.1007/978-3-319-44899-2_15

Nam, T., Pardo, T.A. (2011). Conceptualizing smart city with dimensions of technology, people, and institutions. In: *Proc. 12th Conference on Digital Government Research*, College Park, MD, June 12–15.

Nathansohn, R., Lahat, L. (2022). From urban vitality to urban vitalisation: Trust, distrust, and citizenship regimes in a Smart City initiative. *Cities, 131*, 103969.

O'Grady, M., O'Hare, G. (2012). How smart is your city? *Science, 335*, 1581–1582. https://doi.org/10.1126/science.1217637

Paskaleva, K., Evans, J., Watson, K. (2021). Co-producing smart cities: A Quadruple Helix approach to assessment. *European Urban and Regional Studies, 28*(4), https://doi.org/10.1177/09697764211016037

Patrão, C., Moura, P., Almeida, A.T. (2020). Review of smart city assessment tools. *Smart Cities, 3*, 1117–1132. https://doi.org/10.3390/smartcities3040055

Qayyum, S., Ullah, F., Al-Turjman, F., Mojtahedi, M. (2021). Managing smart cities through six sigma DMADICV method: A reviewbased conceptual framework. *Sustainable Cities and Society, 72*, 103022.

Ramirez Lopez, L.J., Grijalba Castro, A.I. (2021). Sustainability and resilience in smart city planning: A review. *Sustainability, 13*, 181.

Report of the World Commission on Environment and Development Our Common Future, https://sustainabledevelopment.un.org/content/documents/5987our-common-future.pdf, [access data: 9.10.2022].

Rodrigues, H., Bibri, S.E. (2022). *Resilient and Responsible Cities.* Springer: Cham.

Ryba, M. (2017). What is a "smart city" concept and how we should call it in polish, research. *Papers of Wrocław University of Economics, 467*, 82–90.

Samarakkody, A., Amaratunga, D., Haigh, R. (2022). Characterising smartness to make smart cities resilient. *Sustainability, 14*, 12716. https://doi.org/10.3390/su141912716

Sang, Z., Ding, H., Higashi, M., Nakamura, J., Hara, M., Hashitani, T., Sugiura, J., Di Carlo, C., Girdinio, P., Bolla, R. (2015). *Key Performance Indicators Definitions for Smart Sustainable Cities.* ITU-T: Geneva.

Schaffers, H., Komninos, N., Tsarchopoulos, P., Pallot, M., Trousse, B., Posio, E., et al. (2012). Landscape and roadmap of future internet and smart cities. *Technical Report*, hal-00769715f.

Sharifi, A. (2019). A critical review of selected smart city assessment tools and indicator sets. *Journal of Cleaner Production, 233*, 1269–1283.

Shen, L., Huang, Z., Wong, S.W., Liao, S., Lou, Y. (2018). A holistic evaluation of smart city performance in the context of China. *Journal of Cleaner Production, 200*, 667–679.

Somayya, M., Ramaswamy, R. (2016). *Amsterdam Smart City (ASC): Fishing village to sustainable city,* 11 International Conference on Urban Regeneration and Sustainability (SC 2016), https://www.witpress.com/Secure/elibrary/papers/SC16/SC16068FU1.pdf, [access data: 9.10.2022].

Sowa, T. (2022). *Inteligentne miasto – Smart city to nie tylko transport!,* https://mubi.pl/poradniki/inteligentne-miasto/, [access data: 9.10.2022].

Stone, S. (2018). *Key challenges of smart cities & how to overcome them,* https://ubidots.com/blog/the-key-challenges-for-smart-cities/, [access data: 9.10.2022].

The Business Research Company. (2022). *Global Smart Cities Industry.* Global Industry Analysts Inc.: San Jose, CA.

Toli, A.M., Murtagh, N. (2020). The concept of sustainability in smart city definitions. *Frontiers in Built Environment.* https://doi.org/10.3389/fbuil.2020.00077

Townsend, A.M. (2015). *Smart Cities: Big Data, Civic Hackers, and the Quest for a New Utopia.* W. W. Norton & Company: Berlin.

Tratori, R., Rodriguez-Fiscal, P., Pachi, M.A., Koutra, S., Pareja-Eastway, M., Thomas, D. (2021). Unveiling the evolution of innovation ecosystems: An analysis of triple, quadruple, and quintuple helix model innovation systems in European case studies. *Sustainability, 13,* 7582. https://doi.org/10.3390/su13147582

Tripathi, S.L., Ganguli, S., Kumar, A., Magradze, T. (2022). *Intelligent Green Technologies for Sustainable Cities.* Scrivenger Publishing Wiley: New York.

Vershinin, Y.A., Pashenko, F., Olaverri-Monreal, C. (2023). *Technologies for Smart Cities.* Springer: Cham, Switzerland.

Viitanen, J., Kingston, R. (2014). Smart cities and green growth: Outsourcing democratic and environmental resilience to the global technology sector. *Environment and Planning A, 46,* 803–819. https://doi.org/10.1068/a46242

Weir, S. (2015). Are our cities really getting smarter?, *Arqiva.com,* https://www.arqiva.com/news/press-releases/are-our-cities-really-getting-smarter, [access data: 9.10.2022].

Winters, J.V. (2011). Why are smart cities growing? Who moves and who stays. *Journal of Regional Science, 51,* 253–270. https://doi.org/10.1111/j.1467-9787.2010.00693.x

Wolniak, R., Jonek-Kowalska, I. (2022). The creative services sector in Polish cities. *Journal of Open Innovation: Technology, Market, and Complexity, 8*(1), 1–23.

Yoshikawa, Y. (ed.) 2012. *Hitachi's vision of the smart city,* http://www.hitachi.com/rev/pdf/2012/r2012_03_101.pdf, [access data: 9.10.2022].

2 Quality of life and its determinants

2.1 The essence and measurement of the quality of life

Quality of life (QOL) is an interdisciplinary problem of interest to specialists in many areas of science. It is not easy to define the very concept of QOL, which can be understood in different ways. The concept of QOL is still a concept that is sometimes contested in the literature (Prutkin and Feinstein, 2002; Schalock, 2004; Al-Qawasmi, 2019). There is a consensus among researchers that QOL is a multidimensional construct consisting of subjective and objective dimensions covering various aspects of human experience (McCrea et al., 2006; Lora et al., 2010; Von Wirth et al., 2014; Al-Qawasmi, 2020; Wolniak and Jonek-Kowalska, 2021a, 2021b).

The World Health Organization (WHO) defines the QOL as follows: "An individual's perception of his or her position in life in the context of the culture and value systems, in which he or she lives and in relation to his or her goals, expectations, standards and concerns" (WHOQL, 2012). Initially, the WHO definition was the most widely used definition of QOL. In subsequent years, it was expanded to include issues related to how well a person functions in his or her life and how he or she perceives his or her well-being in terms of physical, mental or social aspects of his or her functioning (Hays and Reeve, 2010; Köves et al., 2017; Cai et al., 2021). It can be said that QOL is a concept that encompasses all factors affecting an individual's life.

The concept of urban QOL can be defined as the general well-being of people and societies, who live in cities, along with the quality of the environment, in which they live (Slavuj, 2011; Al-Qawasmi, 2020). It can be said that the quality of urban life consists of both objective attributes (external attributes of the environment and urban space) and subjective attributes (a person's individual observations and perception of tangible and intangible conditions).

DOI: 10.4324/9781003358190-2

A concept derived from the concept of QOL in relation to urban QOL is the concept of urban QOL. It is very important in analyzing issues related to the QOL or urban QOL to take into account subjective perceptions of the phenomenon. Subjective perceptions are important, because many key issues in people's lives, such as the quality of the urban environment, sense of security, sense of social solidarity, sentimental attachment and quality of neighborhood relationships, are difficult to measure using objective indicators (Lora et al., 2010). The literature sometimes criticizes the use of subjective indicators, because of their low reliability, as respondents may differ in their assessment, due to cultural differences, for example (Lora et al., 2010; Al-Qawasmi et al., 2021).

The modern concept of QOL is an interdisciplinary one, influenced by elements such as health, satisfaction of basic material needs (food, clothing, housing), material security (job security, salary), organization of life and work, leisure time that can be spent on personal activities, family and social ties, contacts with nature, education and knowledge, level of independence and personal freedom (Abunazel et al., 2019; Ramirez-Rubio et al., 2019; Gusul and Butnariu, 2021). Given the complexity and multifaceted nature of factors affecting QOL, it is now insufficient to use a single economic indicator, e.g., GDP per capita, as the most important indicator, by which to measure socioeconomic progress. Nowadays, when considering the issue of QOL in the context of sustainable urban development, it is necessary to go beyond the simple measurement of economic values and consider the category of QOL as the most important criterion that can be used in assessing socioeconomic progress (Dawood, 2019). For this to be done, a systematic, holistic approach must be implemented to take into account not only objective factors but also subjective elements, along with the sociocultural context (Rykun et al., 2020; Przybyłowski et al., 2022). To this end, it is necessary to use such methods of measuring the QOL that will ensure the comparability of urban development data through a system of indicators. One of the methods that can be used here is ISO 37120:2018, which will be described in more detail in the next subsection of this publication.

Literature references describe numerous methods of measuring the QOL. Standard indicators used include issues, such as wealth, employment, environment, physical and mental health, education, recreation or leisure time.

One approach to measuring the QOL is to use what is known as The Quality of Life Scale (QOLS), which consists of reliability, validity and utilization. The QOLS was originally an instrument measuring five major areas of QOL, such as material and physical well-being; relationships with other people; social, community and civic activities; personal development and fulfillment; and recreation. The questionnaire was used to measure the QOL as traditionally understood by the WHO definition. In subsequent years, the questionnaire was supplemented with another category – independence, the ability to do things for oneself. Each of the five main categories contained subcategories, which are shown in Table 2.1.

QOL can be measured at both the aggregate and discrete levels. The objective approach to measuring the QOL involves measuring the QOL at the aggregate level by assessing individual physical elements of the environment that contribute to human well-being. Issues, such as: the number or proportion of habitable buildings (homes, schools, hospitals, offices, etc.), infrastructure (e.g., roads, railroads, airports, electricity, sewage and water networks), economic status (GDP, income, employment or assets), environmental status (pollution and climate change) and social services (healthcare, climate change, recreation, education) (Leitmann, 1999; Mohit, 2013). The subjective approach treats QOL as a concept consisting of discrete domains, usually disaggregated at the individual level and more concerned with cognitive experiences, feelings and behavioral dimensions, according to individuals' individual criteria for evaluating and perceiving life (Zayyanu and Abubakr, 2019).

One of the commonly used framework models for improving the QOL in a city is the model proposed by Mitchell, which is shown in Figure 2.1. The model considers the division of QOL into six main areas. Each of the six areas listed can be divided into further components:

1   Health: physical health, mental health;
2   Physical environment: nuisance, climate, pollution, visual perception;
3   Natural resources, goods and services: natural resources, goods, social services;

Table 2.1 Quality of Life Scale

Category	Scale items
Material and physical well-being	Material well-being and financial security
	Health and personal safety
Relationships with other people	Relationships with parents, siblings and other relatives
	Having and raising children
	Relationships with spouses or significant others
	Relationships with friends
Social, community and civic activities	Activities related to helping or encouraging others
	Activities related to local and national governments
Personal development and fulfillment	Intellectual development
	Personal understanding
	Occupational role
	Creativity and personal expression
Recreation	Socializing
	Passive and observational recreational activities
	Active and participatory recreational activities

Source: Burckhardt and Anderson (2003).

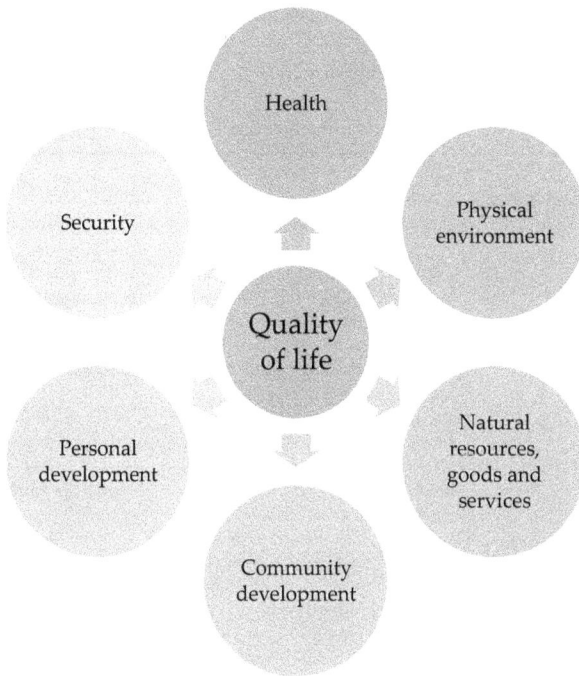

Figure 2.1 The quality of life model.

Source: Mitchell et al. (1995), Zayyanu and Abubakr (2019).

4 Community development: community structure, social network, political participation;
5 Personal development: individual development by learning, individual development by recreation;
6 Security: crime and security, housing, personal economic security.

2.2   Determinants of the quality of life in urban communities

An example of a tool that can be used to measure the QOL is the methodology used in the development of European reports – Report on the QOL in European Cities. For this approach, the following dimensions of urban QOL were used, which are characterized in Table 2.2 (Report, 2020):

• Satisfied with living in the city;
• Safe and inclusive city;
• Getting a job, finding a house and earning a living;
• Moving around the city;

- Culture, squares, parks and healthcare in the city;
- Healthy cities;
- Quality of local public administration.

Table 2.2 Quality of Life in European Cities scale

Category	Characteristics
Satisfied with living in the city	Quality of life depends on aspects that someone else can verify and aspects that only an individual can verify. For example, one's income can be verified, but not whether a person is satisfied with that income (Eurostat, 2016). This is also true for many other issues, such as employment, air pollution, public transportation and safety. Only conducted surveys can reveal people's actual experiences, opinions, feelings and their observations.
	Many quality-of-life issues depend on where people live. From housing costs to clean air, from cultural amenities to transportation, to opportunities, such as access to museums, and risks, such as crime, therefore, where they live (Burger et al., 2020).
Safe and inclusive city	The United Nations 2030 Agenda for Sustainable Development aims to make cities inclusive. The UN has defined an inclusive city as follows: It is a place, where everyone, regardless of economic means, gender, race, ethnicity or religion, has the opportunity and is entitled to participate fully in the social, economic and political opportunities that cities offer (Glatz and Eder, 2019).
	The New Urban Agenda stipulates that cities should prioritize safe, inclusive, accessible, green and high-quality public spaces that are family-friendly, enhance social and intergenerational social interaction, promote social cohesion, inclusion and safety (Hansmaler, 2013).
Getting a job, finding a house and earning a living	Finding a job, a home and earning enough money to live decently is the key to a high quality of life. This includes issues, such as whether it is easy to find a job, find a home and cover expenses.
Moving around the city	Important sites in the city should be accessible to people living in and outside the city.
	Urban transportation can generate problems, such as congestion, traffic accidents, noise and air pollution, as well as greenhouse gases. Consequently, urban transportation networks need to optimize infrastructure use, provide efficient services and encourage a shift to more sustainable modes of transportation (Lättman et al., 2016).
	Transportation in cities should emit less pollution. Achieving sustainable transportation means putting users first and providing them with more affordable, accessible, healthier and cleaner alternatives to their current transportation habits. Furthermore, the Urban Agenda for the EU26 emphasizes that good public transportation is essential for cities and encourages the exchange of best practices between cities.

(Continued)

Table 2.2 (Continued)

Category	Characteristics
Culture, squares, parks and healthcare in the city	Cities often have significant cultural venues, events or programs that can attract large and diverse audiences and contribute to their individual and collective well-being (Blessi et al., 2016; Fancourt and Steptoe, 2018; Grossi et al., 2012; Grossi et al., 2019).
	Cultural and artistic activities can stimulate people's imagination and emotional responses (Ascenso et al., 2021), foster social interaction or healthy lifestyles (Jones et al., 2013), as well as help improve cognitive, creative and relational abilities that improve residents' quality of life and make them feel part of the community (Wilson et al., 2017).
	With a view to promoting cultural participation and its welfare effects, cities should work toward making a wide range of cultural activities available and providing opportunities for active participation in them. In the urban context, green spaces (i.e., parks, public gardens and nearby forests) can play a dual role: on the one hand, they can improve air quality by absorbing pollutants, soaking up rainwater and preventing flooding; on the other hand, they provide opportunities for leisure and sports activities, facilitate social interaction and thus improve the quality of urban life.
Healthy cities	Although air quality has improved over the past decade, air pollution in many European cities exceeds EU air quality standards. Excessive air pollution has a significant negative impact on human health. In addition, long-term exposure to air pollution can have a large negative impact on health. Exposure to PM2.5 is estimated to have caused more than 400,000 premature deaths in 2016 (EEA, 2019).
	Noise pollution is also linked to health problems. An estimated 50 million people in urban areas in Europe are exposed to excessively high levels of traffic noise at night, which can disrupt their sleep. According to the World Health Organization, prolonged exposure to such noise levels can cause elevated blood pressure and heart attacks. The elderly, children and people with poor health are more vulnerable to environmental health risks than the general population (EEA, 2018). In addition, lower socioeconomic status groups (unemployed, people with low incomes or lower levels of education) also tend to be more negatively affected by environmental health hazards, due to their greater exposure and susceptibility, especially in urban areas.
	A city's cleanliness affects its attractiveness and reputation for both residents and visitors. It can also affect residents' assessment of their quality of life, satisfaction with public spaces, their perception of the quality of public services and their overall satisfaction with the city in which they live.

(Continued)

Table 2.2 (Continued)

Category	Characteristics
Quality of local public administration	High-quality public management is associated with higher economic growth, greater impact of public and cohesion policy investments (Rodríguez-Pose and Garcilazo, 2015; European Commission, 2017), higher levels of innovation, less emigration and higher life satisfaction. Also, the quality of governance at the local level varies significantly within the EU (Charron et al., 2010, 2019; European Commission, 2017). Improving the quality of institutions (at all levels of governance) is, therefore, at the heart of the EU and its EU cohesion policy. In the current cohesion policy programming period, 2014–2020, as well as in the upcoming 2021–2027 period, the European Commission is encouraging member states to invest more in capacity building and promoting structural reforms to make the functioning of public administration more efficient and transparent.

Source: Report (2020).

Surveys using the QOL in European Cities scale methodology are conducted annually. Based on the latest research, the following conclusions can be drawn regarding the QOL in European cities (Report, 2020):

- A high level of QOL in northern EU cities and an increase in QOL in eastern EU cities are observed.
- Job satisfaction is high in most cities.
- People feel safer in smaller cities.
- Theft and robbery are more common in large cities and especially in national capitals.
- Cities are seen as a better place for immigrants to live, compared to the rest of the country.
- Most cities are seen as better places to live for the LGBT community, compared to the rest of the country.
- Smaller cities are more elderly friendly.
- Cities outside the countries' capitals are seen as better places to live for young families with children.
- There are problems finding jobs in cities in the south of the EU.
- In most capitals, it is difficult to find a good apartment at a reasonable price.
- In western and northern EU cities, more people are satisfied with the state of their finances.

- Cars are used less frequently in national capitals.
- In larger cities, public transportation is used more intensively.
- Few cities are characterized by high use of bicycles as a means of transportation.
- Satisfaction with public transportation is correlated with its frequent use by residents.
- In order to achieve high satisfaction with public transportation, connections must be frequent.
- Residents of smaller cities are more satisfied with cultural infrastructure.
- The greater the access to green public spaces, the higher the level of resident satisfaction.
- People living in non-capital cities are more satisfied with their public spaces.
- Capital city residents are less satisfied with healthcare.
- More residents are concerned about air quality in cities in the southern and eastern EU.
- Residents consider smaller cities cleaner.
- It is more difficult to follow procedures in public administration in capital cities.
- Online access to city information is easier in northern and western EU cities.
- Perceptions of corruption at the local level vary widely among European Union cities.

ISO 37120 is one of the tools increasingly used in recent years to measure the QOL. ISO 37120:2018 is a solution for measuring the QOL for city services. The standard was published by the International Organization for Standardization in 2018. ISO 37120:2018 focuses on indicators for city services and QOL, offering guidance for city management based on inter-city compatible metrics. It helps cities learn from each other, enabling uniform comparison across a wide range of performance measures, and supports city policy development and prioritization. It is applicable to any city, municipality or local government that wants to measure its performance in a comparable and verifiable way, regardless of size or location.

The standard is designed to enable a uniform assessment of the functioning and achievements of the cities' involvement and is intended to allow a detailed evaluation of their spheres of activity. The criteria used in the standard allow observing and evaluating changes on an annual basis, as well as providing the ability to compare performance with other cities (McCarney, 2015; Komsta, 2016; Fijałkowska and Aldea, 2017).

The standard defines 100 indicators, along with the methodology adopted for their calculation, which can be used by cities of all sizes to measure and control the level of their development from the following points of view: social, economic and environmental (Lehner et al., 2018; Lennova et al., 2018). All indicators have been grouped into 17 thematic areas regarding individual aspects of the city's functioning, which are given as follows (ISO 37120:2018):

1 Economy;
2 Education;
3 Energy;
4 Environment;
5 Finance;
6 Governance,
7 Health,
8 Crisis management;
9 Local government bodies;
10 Recreation;
11 Security;
12 Solid waste;
13 Telecommunication and innovations;
14 Transportation;
15 Urban planning;
16 Wastewater management;
17 Water and wastewater management.

The indicators are divided into 46 primary and 54 secondary indicators. In addition, the standard includes various types of profile indicators that allow cities to decide which ones are most relevant for comparison (Salerno-Kochan, 2016). The indicators included in the standard can be used worldwide by city and business leaders, urban planners, designers, academics and experts to create sustainable, integrated and prosperous cities (Hejduk, 2018; Malinowska and Kurkowska, 2018). The indicators used in ISO 37120:2018 are summarized in Table 2.3. Only core indicators are included in the table.

The standard focuses on three main aspects (Fijałkowska and Aldea, 2017):

1 Transparency in data presentation;
2 Decision accountability;
3 Innovation in becoming a world leader in the implementation of the highest standards of service delivery by the city.

The benefits of ISO 37120:2018 are (ISO 37120:2018; Fijałkowska and Aldea, 2017):

• more efficient management of the city and a higher level of quality in the city's services;

Table 2.3 City core indicators used in ISO 37120:2018

Category	Core indicators
Economy	• City's unemployment rate
Education	• Percentage of female school-aged population enrolled in school
	• Percentage of students completing primary education: survival rate
	• Percentage of students completing secondary education: survival rate
	• Primary education student-teacher ratio
Energy	• Total end-use energy consumption per capita (GJ/year)
	• Percentage of total end-use derived from renewable sources
	• Percentage of city population with authorized electrical service (residential)
	• Number of gas distribution service connections per 100,000 population (residential)
	• Final energy consumption of public buildings per year (GJ/m^2)
Environment and climate change	• Fine particulate matter (PM2.5) concentration
	• Particulate matter (PM10) concentration
	• Greenhouse gas emissions measured in tons per capita
Finance	• Debt service ratio (debt service expenditure as a percentage of a city's own-source revenue)
	• Capital spending as a percentage of total expenditures
Governance	• Women as a percentage of total elected to city-level office
Health	• Average life expectancy
	• Number of in-patient hospital beds per 100,000 population
	• Number of physicians per 100,000 population
	• Under age five mortality per 1,000 live births
Housing	• Percentage of city population living in inadequate housing
	• Percentage of the population living in affordable housing
Population and social conditions	• Percentage of city population living below the international poverty line
Safety	• Number of firefighters per 100,000 population
	• Number of fire-related deaths per 100,000 population
	• Number of natural-hazard-related deaths per 100,000 population
	• Number of police officers per 100,000 population
	• Number of homicides per 100,000 population
Solid waste	• Percentage of city population with regular solid waste collection (residential)
	• Total collected municipal solid waste per capita
	• Percentage of the city's solid waste that is recycled
	• Percentage of the city's solid waste that is disposed of in a sanitary landfill
	• Percentage of the city's solid waste that is treated in waste-to-energy plants
Sports and culture	• Number of cultural institutions and sporting facilities per 100,000 population
Transportation	• Kilometers of public transport system per 100,000 population
	• Annual number of public transport trips per capita

(Continued)

Table 2.3 (Continued)

Category	Core indicators
Urban/local agriculture and food security	• Total urban agricultural area per • 100,000 population
Urban planning	• Green area (hectares) per 100,000 population
Wastewater	• Percentage of city population served by wastewater collection • Percentage of city's wastewater receiving centralized treatment • Percentage of population with access to improved sanitation
Water	• Percentage of city population with potable water supply service • Percentage of city population with sustainable access to an improved water source • Total domestic water consumption per capita (liters/day) • Compliance rate of drinking water quality

Source: ISO 37120:2018.

- provision of a framework for sustainable development and strategic planning of the city;
- obtaining international targets and benchmarks for the analyses to be carried out;
- benchmarking and local planning;
- information for city managers and decision-makers used in decision-making;
- access to data, including the possibility of increasing the reliability of these data through auditing and verification by external institutions;
- urban learning;
- greater credibility in the financial markets, greater chances of attracting investors and introduction of financing programs for activities;
- obtaining a sustainable development planning framework;
- transparency and openness of data for investment attractiveness;
- comparability of data on the city's decisions, its appearance and global benchmarking;
- the usefulness of the certificate in efforts to obtain city funding from EU funds.

In order to use the standard in cities to report a different range of data, a set of certification levels was used for the standard, which depend on how many of the listed indicators are monitored in a given city. Synthetically, the different certification levels of the standard are shown in Table 2.4.

The main advantage of the ISO 37120 standard comes not only from basing the assessment on a set of indicators but also from the possibility of comparing the results obtained and benchmarking the data between different cities undergoing the certification in question (Wang and Fox,

Table 2.4 Implementation levels of ISO 37120

Level	Characteristics
Aspiring	30–45 primary indicators
Bronze	46–59 indicators (46 primary and 0–13 secondary)
Silver	60–75 indicators (46 primary and 14–29 secondary)
Gold	76–90 indicators (46 primary and 30–44 secondary)
Platinum	91–100 indicators (46 primary and 45–54 secondary)

Source: ISO 37120:2018.

2017; McCarney, 2015;). The main problem with collecting this type of data is that it can be communicated between cities and that access is open. To ensure this, the World Council on City Data (WCCD) open data platform, based in Toronto, was opened (Kowalczyk, 2018). This platform coordinates all activities that are related to city data reported according to the ISO 37120 standard, as well as other normative solutions based on this standard. Individual cities that have obtained ISO 37120 certification are added to the organization's Global Cities Registry™ database for a period of one year. At the end of this period, they must go through the certification process again. All data reported by cities are posted on a specially developed virtual platform http://open.dataforcities.org/, which provides open access to them.

Furthermore, the World Council on City Data has set numerous goals that cities should successively achieve by monitoring and improving individual QOL indicators. Some of the most important goals are (McCarney, 2015) given as follows:

- Poverty eradication;
- Eradication of hunger;
- Good health of residents;
- Quality education;
- Ensuring gender equality;
- Access to clean water and cleaning supplies;
- Clean and cheap energy;
- Decent working conditions and economic growth;
- Industry, innovation and infrastructure;
- Inequality reduction;
- Promoting the idea of a sustainable city and a sustainable society;
- Responsible consumption and production;
- The fight against climate change;
- Sustainable use of water resources;
- Sustainable use of ecosystems on land;
- Peace, justice and strong institutions;
- Partnership to achieve the listed goals.

2.3 Problems and challenges of improving the quality of life in contemporary cities

Analyzing different approaches to urban QOL and measuring it, it is important to note the different types of problems and challenges modern societies have to face, which should be taken into account when planning a modern Smart City, in order to achieve a higher level of QOL for residents in this type of city.

Based on the analysis of the results of studies conducted in the field of QOL in European Union countries, the following main problems can be distinguished, which should be solved to improve the QOL of the population (The Future, 2019; Vardoulakis and Kinney, 2019; Mourtadis, 2021):

- Affordable housing – Some of the cities in Europe, where new residents are most likely to settle, have seen housing prices soar in recent years. This threatens housing affordability, as prices are rising faster than incomes and the availability of new housing is low. The recent increase in foreign and corporate investment in urban residential real estate has led to changing ownership patterns, raising concerns about the social fabric of the city and who can own the city and who can be responsible for citizens' rights to affordable housing. Short-term rental platforms, which are becoming increasingly popular, may cause excessive increases in real estate prices and negatively affect local QOL.
- Mobility – Urban mobility is one of the areas that will undergo the greatest changes in the future as a result of technological innovation and changing behavior of residents. The number of private vehicle owners is likely to decline, as mobility understood as a service, combining multiple modes of transportation, becomes increasingly popular in cities. Legislation and appropriate management measures will need to be adapted to this phenomenon to ensure that the new modes of transportation complement traditional public transportation, rather than compete with it. Autonomous electric vehicles can benefit cities by reducing air pollution and traffic congestion.
- Provision of services – In the future, specialized urban services, an essential element of a city, should be planned in a sustainable, efficient, reusable, sharable, modular, personalized and data-driven manner. The nature of public and commercial services in cities is constantly transforming. Specialized (regional) services require a large nearby market and are, therefore, more profitable in larger cities. Service delivery can be improved by promoting compact urban development, developing integrated land use and mobility plans, and using new technologies to facilitate service delivery.
- Aging – By 2070, life expectancy in the EU will rise to 88.2 years, and the old-age dependency ratio (the number of older people relative to

the number of people of working age) is expected to decline. While population aging is a global trend, it is of particular concern in regions, where the total population is declining, as is increasingly the case in Europe. An additional burden will be placed on the social welfare system, as rising costs of healthcare, pensions and social benefits will have to be met by a shrinking workforce, which could affect overall GDP and innovation. Cities will need to adjust their services in areas such as healthcare and mobility, as well as public infrastructure, housing and social policies to accommodate changing demographics.

- Urban health – Health outcomes can be improved by changing the urban structure of cities and towns: urban planning plays a crucial role in achieving health improvements.
- Social segregation – Integrated anti-segregation policies should take into account the diverse factors present in poor neighborhoods (e.g., health, housing and ethnicity). Urban policies that promote diversity can become drivers of innovation.
- Environmental footprint – Resource consumption affects not only local but also global sustainability. Providing water, energy and food security for urban populations puts significant pressure on the environment beyond city limits. Although water consumption in most economic sectors in Europe has declined since 1990, water availability problems are expected to increase. Lifestyle and behavioral changes can help urban residents significantly reduce their environmental footprint, such as switching to a healthy diet, reducing waste, using active or public transportation, or choosing sustainable energy sources.
- Climate action – Cities are responsible for high levels of energy consumption, and thus, generating about 70% of global greenhouse gas emissions, cities are particularly vulnerable to the effects of climate change. Cities are most effective in taking action to address climate change when they are connected to each other and to actors at the national and regional levels.
- Data availability and management.
- Management of emerging technologies.
- The changing role of society.
- Integrated policy design.

Notably, many of the problems related to the QOL in cities can be solved by introducing Smart Cities solutions. Several examples can be cited here (Woetzel et al., 2018; Li et al., 2019; 5 Ways, 2020):

- Improving public safety – The use of applications that could potentially reduce fatalities in homicides, fires and traffic accidents by 8–10%. Incidents of muggings, burglaries, car thefts and robberies could then be reduced by 30–40%. When it comes to crime, cities can use data to make more efficient use of limited resources and personnel. For example, real-time crime mapping uses statistical analysis to find patterns of behavior and improve public safety. Predictive policing can anticipate crimes, and when they occur, applications such as home security systems, gunshot detection and smart surveillance can make law enforcement respond faster.

- Speeding up daily commutes – One aspect critical to improving QOL is improving residents' daily commutes. By 2025, cities that implement smart mobility applications can reduce commute times by an average of 15–20%. This is related to variables, such as the city's population density, commuting patterns and existing transit infrastructure. Installing IoT sensors on existing physical infrastructure can help solve traffic problems before they turn into breakdowns and delays. Applications that alleviate traffic congestion are most effective in cities, where driving and buses are the main forms of transportation. Smart synchronization of traffic signals can potentially reduce average commutes by more than 5% in developing cities, where many people travel by bus. Real-time navigation alerts inform drivers of delays, helping them choose the fastest route, while smart parking apps direct them to available spots.

- Better public health – Apps that help monitor, prevent and treat chronic diseases, such as diabetes and cardiovascular disease, have the greatest potential to improve the situation in developed countries. Remote patient monitoring systems may potentially reduce the health burden in high-income cities by more than 4%. These systems use digital devices to take key readings and then send them to doctors for evaluation. With these data, the patient and doctor would know if early intervention is needed, reducing complications and hospitalizations. Cities can also use the data and analysis to identify demographic groups that have a higher risk profile, allowing for more targeted medical interventions. If developing cities use infectious disease surveillance systems to stay ahead of fast-moving epidemics, a 5% reduction in cases is possible. Finally, telemedicine, which is a growing trend, particularly during the COVID-19 pandemic, could be a life-saving measure used in low-income cities that lack doctors.

- Cleaner and more sustainable environment – With the growth of urbanization, industrialization and consumption, human pressure on the environment is increasing. Applications, such as building automation systems, some mobile applications and dynamic analysis of electricity consumption, can help reduce emissions by 10–15%. Water consumption tracking combines advanced metering systems with digital feedback. This can encourage people to conserve and reduce consumption by 15% in cities, where residential water use is high. The biggest source of water waste in developing countries is water leakage from pipes. The use of sensors and analytics can reduce these losses by up to

25%. Applications, such as digital tracking, can reduce solid waste per capita by 10–20%. Through the use of smart apps, cities can reduce the amount of unrecycled solid waste by 30–130 kilograms per person per year and save 25–80 liters of water per person each day. As for air health, air quality sensors can identify sources of pollution and provide a basis for further action. Also, making real-time air quality information available to the public through smartphone apps allows individuals to take protective measures. Depending on the current level of pollution, this can reduce negative health effects by 3–15%.

- Strengthening social ties – McKinsey's analysis found that using applications such as digital channels to communicate with local officials, as well as digital platforms that lead to real-world interactions (such as Nextdoor and Meetup) can nearly triple the percentage of residents, who feel connected to their local government and double the percentage of those, who feel connected to their local community. Creating channels for two-way communication between local authorities and the public can make city governments faster and more responsive to residents' needs. Many city organizations now have an active presence on social networks, and others have created their own interactive applications for citizens. In addition to disseminating information, these channels create platforms for residents to collect data, report problems or express opinions on planning issues.

The challenges facing modern cities in terms of QOL depend, to some extent, on the geographic region. Each region has its own specificity, and consequently, slightly different challenges may be faced by city governments in terms of the factors that affect the improvement or deterioration of QOL. For example, in the case of Latin America, the following challenges can be identified to overcome to ensure that the QOL in cities improves (Liberlun, 2021):

- Structural social exclusion – Urban inequality runs persistent and deep, with major cities in many countries experiencing greater inequality than the country as a whole, and inequality in some cities increasing as the number of people living in poverty decreases (UN Habitat, 2016). The likelihood of living in a neighborhood with inadequate public services depends largely on ethnicity, place of birth and other characteristics beyond people's control. Latin American and Caribbean cities are underprovided with safe public green spaces, and their distribution and quality are uneven. Gaps in urban service provision particularly affect women, children, the elderly and the disabled, who make up about two-thirds of the city's population.
- Excessive pollution and poor mitigation and resilience to negative climate change – Cities can reduce toxic emissions and improve QOL by transforming their urban planning, urban environment and better

energy use. The region has made some progress in introducing energy- and water-efficient technologies in housing, but much remains to be done to reduce the environmental footprint of cities. Cities also have high levels of noise pollution, causing health problems and depressing real estate prices. Latin American and Caribbean cities are also highly vulnerable to disasters caused by natural risks and climate change.

- Stagnating urban productivity – Latin American and Caribbean productivity depends heavily on a few cities, creating the risk that economic shocks in these cities could destabilize the entire region's economy. Poor infrastructure between and within cities undermines productivity. Also, burdensome city regulations increase costs for small entrepreneurs and contribute to the persistence of lack of employment formality. Also, Latin American cities are failing to take full advantage of the opportunities that innovation in the built environment provides for increasing urban productivity.
- Weak urban governance – Latin America's urban governance institutions have limited capacity to address the complex and interdisciplinary problems they face. Most city governments have limited fiscal autonomy, insufficient financial and human resources, and little access to data and technology. Municipal governments are still lagging behind in using digital technologies to engage in open dialog with residents.

Bibliography

5 Ways Smart Cities Improve the Urban Quality of Life. (2020). https://stefanini. com/en/insights/news/5-ways-smart-cities-improve-the-urban-quality-of-life, [access data: 3.10.2022].

Abunazel, A., Hammad, Y., Abd AlAziz, M., Gouda, E., Anis, W. (2019). Quality of life indicators in sustainable urban areas. *The Academic Research Community publication, 3*, 78.

Al-Qawasmi, J. (2019). Exploring indicators coverage practices in measuring urban quality of life. *Proceedings of the Institution of Civil Engineers: Urban Design and Planning, 172*(1), 26–40. htps://doi.org/10.1680/jurdp.18.00050

Al-Qawasmi, J. (2020). Measuring quality of life in urban areas: toward an integrated approach. *Journal of Environmental Science and Natural Resources, 25*(2), 67–74. htps://doi.org/10.19080/IJESNR.2020.25.556158

Al-Qawasmi, S.M., Asfour, O., Aldosary, A.S. (2021). Assessing urban quality of life: Developing the criteria for Saudi cities. *Frontier in Built Environment.* htps://doi.org/10.3389/fbuil.2021.682391

Ascenso, A., Augusto, B., Silveira, C., Roebeling, P., Miranda, A.I. (2021). Impacts of nature-based solutions on the urban atmospheric environment: A case study for Eindhoven, The Netherlands, *Urban Forestry and Urban Greening, 57*, 126870.

Blessi, G.T., Grossi, E., Sacco, P.L., Pieretti, G., Ferilli, G. (2016). The contribution of cultural participation to urban well-being. A comparative study in Bolzano/Bozen and Siracusa, Italy. *Cities, 50*, 216–226.

Burckhardt, C.S., Anderson, K.L. (2003). The quality of life scale (QOLS): Reliability, validity, and utilization. *Health and Quality of Life Outcomes, 1*, 60–70. htps://doi.org/10.1186/1477-7525-1-60

Burger, M.J., Morrison, P.H., Hendriks, M., Hoogerbrugge, M.M. (2020). Urban-rural happiness differentials across the world. In: Helliwell, J.F., Layard, R., Sachs, J., De Neve, J.E. (eds) *World Happiness Report 2020.* Sustainable Development Solutions Network: New York.

Cai, T., Verze, P., Truls, E., Johansen, B. (2021). The quality of life definition: Where are we going? *Uro Journal of Urology, 1*, 14–22. htps://doi.org/10.3390/uro1010003

Charron, N., Dijkstra, L., Lapuente, V. (2010). Regional governance matters: Quality of government within European Union Member States. *Regional Studies, 48*(1), 68–90.

Charron, N., Lapuente, V., Annoni, P. (2019). Measuring quality of government in EU regions across space and time. *Papers in Regional Science, 98*(5), 1925–1953.

Dawood, S.R.S. (2019). Sustainability, quality of life and challenges in an emerging city region of George Town, Malaysia. *Journal of Sustainable Development, 12*, 35.

European Commission. (2017). *Seventh report on economic, social and territorial cohesion.* Publications Office of the European Union: Luxembourg.

European Environmental Agency. (2018). *Unequal exposure and unequal impacts: Social vulnerability to air pollution, noise and extreme temperatures in Europe*, EEA Report No 22/2018. Publications Office of the European Union: Luxembourg.

European Environmental Agency. (2019). *Air quality in Europe – 2019 Report*, EEA Report No 10/2019. Publications Office of the European Union: Luxembourg.

Eurostat. (2016). *Analytical report on subjective well-being*, 2016 edition. Publications Office of the European Union: Luxembourg.

Fancourt, D., Steptoe, A. (2018). Community group membership and multidimensional subjective well-being in older age. *Journal of Epidemiology and Community Health, 72*(5), 376–382.

Fijałkowska, J., Aldea, T. (2017). Raportowanie zrównoważonego rozwoju miasta. Norma 37120, *Prace Naukowe Uniwersytetu Ekonomicznego we Wrocławiu, 478*, 174–183.

Glatz, C., Eder, A. (2019). Patterns of trust and subjective well-being across Europe: New insights from repeated cross-sectional analyses based on the European Social Survey 2002–2016. *Social Indicators Research, 148*, 417–439.

Grossi, E., Tavano Blessi, G., Sacco, P.L., Buscema, M., Blessi, G.T., Sacco, P.L., Buscema, M. (2012). The interaction between culture, health and psychological well-being: Data mining from the Italian culture and well-being project. *Journal of Happiness Studies, 13*(1), 129–148.

Grossi, E., Tavano Blessi, G., Sacco, P.L. (2019). Magic moments: Determinants of stress relief and subjective wellbeing from visiting a cultural heritage site. *Culture, Medicine and Psychiatry, 43*(1), 4–24.

Gusul, P.F., Butnariu, A.R. (2021). Exploring the relationship between smart city, sustainable development and innovation as a model for urban economic growth. *Annals of the University of Oradea: Economic Science, 30*, 82–91.

Hajduk, S. (2018). The smartness profile of selected European cities in urban management – A comparison analysis. *Journal of Business Economics and Management, 19*(6), 797–812.

Hanslmaier, M. (2013). Crime, fear and subjective well-being: How victimization and street crime affect fear and life satisfaction. *European Journal of Criminology, 10*(5), 515–533.

Hays, R.D., Reeve, B.B. (2010). Measurement and modeling of health-related quality of life. In: Killewo, J., Heggenhougen, H.K., Quah, S.R. (eds) *Epidemiology and Demography in Public Health*. Academic Press: San Diego, CA, 195–205.

ISO 37120:2018, Sustainable cities and communities—Indicators for city services and quality of life, 2018.

Jones, M., Kimberlee, R., Deave, T., Evans, S. (2013). The role of community centre-based arts, leisure and social activities in promoting adult well-being and healthy lifestyles. *International Journal of Environmental Research and Public Health, 10*(5), 1948–1962.

Komsta, H. (2016). Rankingi jako narzędzia oceny Inteligentnego Miasta. *Logistyka, 1,* 22–24.

Köves, B., Cai, T., Veeratterapillay, R., Pickard, R., Seisen, T., Lam, T.B., Yuan, C.Y., Bruyere, F., Wagenlehner, F., Bartoletti, R. (2017). Benefits and harms of treatment of asymptomatic bacteriuria: A systematic review and meta-analysis by the European Association of Urology Urological Infection Guidelines. *Panel European Urology, 72,* 865–868.

Kowalczyk, A. (2018). Normalizacja standardów w administracji samorządowej. *Przegląd Komunalny, 11,* 79–81.

Lättman, K., Margareta, F., Olsson, L.E. (2016). Perceived accessibility of public transport as a potential indicator of social inclusion. *Social Inclusion, 3*(4), 36–45.

Lehner, A., Erlacher, C., Schlögl, M., Wegerer, J., Blaschke, T., Steinnocher, K. (2018). Can ISO-defined urban sustainability indicators be derived from remote sensing: An expert weighting approach, *Sustainability, 10*(4), 1268.

Leitmann, J. (1999). Can city QOL indicators be objective and relevant? Towards a participatory tool for sustaining urban development. *Local Environment, 4*(2), 169–180.

Lennova, T., Golovcova, I., Mamedov, E., Varfolomeeva, M. (2018). *The integrated indicator of sustainable urban development based on standardization.* International Science Conference on Business Technologies for Sustainable Urban, Article number 01039.

Li, D., Cheng, T., Genderen, J.L., Shao, Z. (2019). Challenges and opportunities for the development of MEGACITIES. *International Journal of Digital Earth, 12*(12), 1382–1395. doi: 10.1080/17538947.2018.1512662

Liberlun, N. (2021). *Four challenges our cities can overcome,* https://blogs.iadb. org/ciudades-sostenibles/en/four-challenges-our-cities-can-overcome/, [access data: 3.10.2022].

Lora, E., Powell, A., van Praag, B., Sanguinetti, P. (2010). Preface. In: Lora, E., Powell, A., van Praag, B.M.S., Sanguinetti, P. (eds) *The Quality of Life in Latin American Cities: Markets and Perception*. The Inter-American Development Bank: Washington, DC, xix–xxiv.

Malinowska, E., Kurkowska, A. (2018). Norma ISO 37120 narzędziem pomiary idei zrównoważonego Rozwoju miast, *Zeszyty Naukowe Politechniki Śląskiej, seria: Organizacja i Zarządzanie, 118*, 363–382.

McCarney, P. (2015). The evolution of global city indicators and ISO 37120: The first International standard on city indicators. *Statistical Journal of the IAOS, 31*, 131–142.

McCrea, R., Shyy, T.-K., Stimson, R. (2006). What is the strength of the link between objective and subjective indicators of urban quality of life? *Applied Research in Quality of Life, 1*, 79–96. htps://doi.org/10.1007/s11482-006-9002-2

Mitchell, G., May, A., McDonald, A. (1995). PICABUE: A methodological framework for the development of indicators of sustainable development. *International Journal of Sustainable Development & World Ecology, 2*(2), 104–123.

Mohit, M.A. (2013). Objective analysis of variation in the regional quality of life in Malaysia and its policy implications. *Procedia – Social and Behavioral Sciences, 101*, 454–464.

Mourtadis, K. (2021). Urban planning and quality of life: A review of pathways linking the built environment to subjective well-being. *Cities, 115*, 103229. https://doi.org/10.1016/j.cities.2021.103229

Patel, U., Rakshit, S., Ram, S.A., Irfan, Z.B. (2019). Urban sustainability index: Measuring performance of 15 metropolitan cities of India. *South Asian Journal of Social Studies and Economics, 3*, 1–11.

Prutkin, J.M., Feinstein, A.R. (2002). *A History of Quality of Life Measurements.* Yale Medicine Thesis Digital Library, 424. Available at: http://elischolar.library.yale.edu/ymtdl/424.

Przybyłowski, A., Kałaska, A., Przybyłowski, P. (2022). Quest for a tool measuring urban quality of life: ISO 37120 standard sustainable development indicators. *Energies, 15*, 2841. htps://doi.org/10.3390/en15082841

Ramirez-Rubio, O., Daher, C., Fanjul, G., Gascon, M., Mueller, N., Pajín, L., Plasencia, A., Rojas-Rueda, D., Thondoo, M., Nieuwenhuijsen, M. (2019). Urban health: An example of a 'health in all policies' approach in the context of SDGs implementation. *Global Health, 15*, 87.

Report on the Quality of Life in European Cities. (2020). *Luxembourg: Publication Office of the European Union.* https://ec.europa.eu/regional_policy/sources/docgener/work/qol2020/quality_life_european_cities_en.pdf, [access data: 3.10.2022].

Rodríguez-Pose, A. (2013). Do institutions matter for regional development? *Regional Studies, 47*(7), 1034–1047.

Rodríguez-Pose, A., Garcilazo, E. (2015). Quality of government and the returns of investment: Examining the impact of cohesion expenditure in European regions, *Regional Studies, 49*(8), 1274–1290.

Rykun, A.Y., Chernikova, D.V., Sukhushina, E.V., Beryozkin, A.Y. (2020). Measuring the quality of life in Urban areas: The feasibility of using the index approach. *Zhurnal Issled. Sotsial'noi Polit, 18*, 283–298.

Salerno-Kochan, M. (2016). Norma ISO 37120. *Próba oceny jakości życia w miastach,* [w:] M. Salerno-Kochan (red.), Wybrane aspekty zarządzania jakością. Kraków: Polskie Towarzystwo Towaroznawcze.

Schalock, R.L. (2004). The Concept of Quality of Life: What We Know and Do Not Know. *Journal of Intellectual Disability Research, 48*(3), 203–216. htps:// doi.org/10.1111/j.1365-2788.2003.00558.x

Slavuj, L. (2011). Urban quality of life - a case study: The city of rijeka | Kvaliteta života – studija slučaja: Grad Rijeka, *Hrvatski Geografski Glasnik, 73*(1), 99–100.

The future of Cities. Opportunities, Challenges and the way forward. (2019). European Commission, Luxembourg, Publications of European Union, https:// www.google.com/url?sa=t&rct=j&q=&esrc=s&source=web&cd=&cad=rja&ua ct=8&ved=2ahUKEwjXiZOv2sT6AhVokIsKHQN4CogQFnoECAUQAQ&ur l=https%3A%2F%2Fpublications.jrc.ec.europa.eu%2Frepository%2Fbitstream%2 FJRC116711%2Fthe-future-of-cities_online.pdf&usg=AOvVaw33of1EEL09bs QOwDHW0qcE, [access data: 3.10.2022].

UN Habitat. (2016). *Urbanization and development. Emerging Futures, Word Cities report,* Nairobi, Kenya, https://unhabitat.org/sites/default/files/ download-manager-files/WCR-2016-WEB.pdf, Nairobi, Kenya, [access data: 3.10.2022].

Vardoulakis, S., Kinney, P. (2019). *Grand Challenges in Sustainable Cities and Health, Frontiers in Sustainable Cities.* htps://doi.org/10.3389/ frsc.2019.00007

Von Wirth, T., Hayek, U.W., kunze, A., Neuenschwander, N., Stauffacher, M., Scholz, R.W. (2014). Identifying urban transformation dynamics: Functional use of scenario techniques to integrate knowledge from science and practice, *Technological Forecasting & Social Change, 89,* 115–130. https://doi. org/10.1016/j.techfore.2013.08.030

Wang, Y., Fox, M.S. (2017). *Consistency analysis of city indicator data,* 15th International Conference on Computers in Urban Planning and Urban Management, 355–369.

WHOQOL: Measuring Quality of Life. World Health Organization (2012), https://www.who.int/toolkits/whoqol, [access data: 3.10.2022].

Wilson, N., Gross, J., Bull, A. (2017). *Towards cultural democracy: Promoting cultural capabilities for everyone,* https://www.kcl.ac.uk/Cultural/culturalenquir ies/Towards-cultural-emocracy/Towards-Cultural-Democracy-2017-KCL.pdf, [access data: 3.10.2022].

Woetzel, J., Remes, J., Boland, B., Lv, K., Sinha, S., Strube, G., Means, J., Law, J., Cadena, A., Tann, V. (2018). *Smart cities: Digital solutions for a more livable future,* McKinsey Global Institute, https://www.mckinsey.com/capabilities/ operations/our-insights/smart-cities-digital-solutions-for-a-more-livable-future#part1, [access data: 3.10.2022].

Wolniak, R., Jonek-Kowalska, I. (2021a). The level of the quality of life in the city and its monitoring. *Innovation (Abingdon), 34*(3), 376–398.

Wolniak, R., Jonek-Kowalska, I. (2021b). The quality of service to residents by public administration on the example of municipal offices in Poland. *Administration Management Public, 37,* 132–150.

Zayannu, M., Abubakr, I.R. (2019). *Improving the quality of life of urban communities in developing countries.* In: Leal Filho et al. (eds) *Responsible Consumption and Production, Encyclopedia of the UN Sustainable Development Goals.* Springer: Cham. htps://doi.org/10.1007/978-3-319-71062-4_25-1

3 Smart City Governance

3.1 The role of city authorities in creating smart cities

The Smart City concept can fulfill a beneficial and desirable role in spatial city management. Spatial management in urban units to incorporate Smart City aspects should be based on knowledge and innovation. Smart City specialists (Hollands, 2008; Caragliu et al., 2011; Dameri, 2013; Komninos, 2014; Vujković et al., 2022) emphasize the need to invest in human and social capital and the use of information technology in order to improve the quality of life in the city and implement the principles of sustainable development in the management process. A city managed according to the guidelines of the Smart City concept should effectively solve social and environmental problems (Hajduk, 2016). This type of city has all the components of an urban settlement unit thanks to the construction of smart connections of the activities of self-determined, independent and informed citizens (Albino et al., 2015).

Related to the Smart City concept is the concept of Smart Governance. Smart Governance is public management in which issues such as public participation in decision-making, transparency of operations, quality and accessibility of public services play an important role (Faraji et al., 2021; Carrato-Gómez and Roig-Segovia, 2022; Bokhari and Meyong, 2022). Real-time management using modern technologies and smart infrastructure networks is very important for Smart Governance (He et al., 2022; Nina et al., 2022).

Smart City Governance is defined as the analysis of comprehensive activities and services in the public domain of a smart city. Smart City Governance is extremely important for the management of smart cities (Lim and Yigitcanlar, 2022; Willis and Nold, 2022). It should be oriented toward making citizens very well informed about the operation of their smart city and transparency of the entire system (Vitálišová et al., 2022). Table 3.1 presents the most important arguments for the importance of implementing Smart Governance.

DOI: 10.4324/9781003358190-3

Table 3.1 Important aspects of Smart City Governance

Aspect	Characteristics
Determining tangible benefits to citizens	Before investing in Smart City technology and digital infrastructure, local government bodies must first consider and evaluate the potential tangible benefits that smart cities can provide to their citizens.
	Therefore, it is important to define the difference between the needs of citizens and their desires. Furthermore, local authorities must ensure that the basic needs of citizens are not compromised.
Explaining the intended purpose of technology and funding	Implementing the Smart City concept can inherently change citizens' lives. Therefore, it is crucial for a smart city management to explain in detail to citizens the intended purpose of the Smart City technology being implemented.
	Moreover, smart cities will only be able to achieve optimal Smart City technology solutions with the help of local resident involvement and financial investment. Therefore, Smart City solutions should always focus on those local residents who will be most affected by the change.
Data security and encryption concerns	The technology used to implement smart city solutions is usually heavily reliant on monitoring user behavior patterns, sharing data and gaining insights into various consumer behaviors. Some citizens may consider these practices an invasion of privacy.
	Therefore, it is extremely important for local government bodies to explain the benefits of Smart City technology. It should be explained to citizens how their personal data and information will be protected and encrypted. In addition, it is also extremely important that the data collected be used for predictive analysis and in terms of public discourse. Otherwise, data tracking may be perceived as unnecessarily invasive.
Consumer trust	Even with sufficient public investment and funding, Smart City technology will not easily achieve maximum return on investment until consumers trust the digital technology provided. Local governments need to clearly communicate and demonstrate to local citizens the importance, use, benefits and positive effects of Smart City technology. This can be achieved through programs to improve public awareness of the issue.

Source: Arora (2021).

The implementation of the Smart City concept at the city level, along with a description of the role of city authorities at its various stages, is included in the Smart City Maturity Model. The model is an attempt to measure the level of city maturity, which is an important task facing city authorities, and it also gives guidance on the actions that city authorities should take at the next stages of implementing the Smart City concept (Hajduk, 2020). The Smart City Maturity Model concept defines the basic stages, dimensions, results and actions that should be taken by city authorities. The use of the model makes it possible to assess the competence and identify the shortcomings of a city in achieving a higher level of Smart City maturity. The model serves to improve data-driven decision-making and achieve desired financial, social and environmental outcomes related to city-wide goals.

The Smart City management model includes five basic stages in the use and analysis of data, such as ad hoc, opportunistic, iterative, managerial and optimized. Table 3.2 includes the listed five stages of Smart City management development along with their characteristics.

For city governments to implement Smart City concepts, they should focus on implementing the concept according to the stages mentioned. It is worth noting that currently, the pace of digital transformation is tremendous, and data, population potential and technology should be used by cities to achieve smart change (Waarts, 2016; Ependi et al., 2022). Having a proper strategy for implementing Smart City concept and smart city management and implementing best practices in this area helps reduce the complexity and risks associated with many aspects of smart city operations such as initiatives in smart transportation, smart parking systems, environmental monitoring and lighting solutions (Mohsin et al., 2019; Shi and Cao, 2022).

Table 3.3 presents the characteristics of the various dimensions of the implementation of the Smart City concept in the city based on the Smart City Maturity Model concept. The Smart City Maturity Model is closely related to the city development strategy and the inclusion of the Smart City concept in it. Detailed information and case studies on Smart Cities strategies in Central and Eastern Europe cities will be included in the next subsection of this chapter.

The Sustainable Smart Cities Maturity Model defines the levels, key measures, targets and actions that are recommended for cities to successfully move through the levels and examine their current situation and

Table 3.2 Stages of development of Smart City management

Stage	Characteristics	Objective	Results
Ad hoc	Closed	Provision of services	Technology-related successes
Opportunistic	Cooperation	Stakeholder participation	Management basics and strategic planning
Iterative	Integrated	Improved results	Rationalization
Managerial	Operationalization	Forecasting and prevention	Adaptive response system
Optimized	Sustainable development	Competitive differentiation	Innovation and continuous improvement

Source: Hajduk (2020), Clark (2017).

Table 3.3 Characteristics of the dimensions of Smart City implementation in the city

Dimensions	Characteristics
Strategic objective	An effective Smart City has a strategy and action plan that identifies how investments in data and digital technologies enable service reform and partnerships. An effective strategy focuses on delivering better results in line with the city's strategic priorities.
Data	A successful Smart City effectively leverages its data assets to deliver better results. It is investing in a system that includes data capture, integration and analysis capabilities. Open data underpin their commitment to transparency and innovation.
Technology	An effective Smart City invests in open, flexible, integrated and scalable ICT architectures that enable accelerated service innovations, such as providing automated and dynamic responses in real time.
Management and service delivery models	An effective Smart City adapts traditional organizational models of delivery to leverage the power of data and digital technologies. It invests in systemic partnership models focused on shared outcomes
Stakeholder engagement	An effective Smart City makes the best use of data and digital technologies to invest in increased openness and transparency. It proactively improves the uptake of digital services while supporting the digitally excluded.

Source: Hajduk (2020), Overview (2016), Smart (2019).

identify the critical capabilities needed to achieve the long-term goal of becoming a Sustainable Smart City. The objective of maturity models, such as the Sustainable Smart Cities model, is to help cities and related stakeholders improve intra- and inter-city cooperation in defining and implementing city development strategies and promoting and encouraging the use of new technologies and solutions (Founoun et al., 2022). Using the model can improve smart city management. Table 3.4 provides highlights of selected aspects of smart city management that should be implemented at successive maturity levels of the Sustainable Smart City Maturity Model (Smart, 2019).

Cities that adopt a Smart City strategy make city services more efficient and make them more attractive to investors, residents, visitors and businesses (Laurini, 2021; Maurya and Biswas, 2021). These cities benefit from working together on this by identifying common approaches and solutions that are transferable to other cities. By working together, cities can position themselves to access investment, accelerate progress through learning and identify areas of local innovation that can be scaled.

The implementation of the Smart City concept requires the broad participation of local authorities, which can improve the management of many areas of a smart city in this way (Fonseca et al., 2021; Saadah, 2021;

Table 3.4 Smart City maturity levels in the Sustainable Smart Cities Maturity Model

Maturity level	Characteristics
Level 1	At this level, the main goal the city needs to achieve is to have a municipal Smart City strategy with an associated roadmap. A clear roadmap or strategic plan should be used to develop a Smart City based on ICT.
	Achievements at this level may include the following issues:
	• The city has developed a detailed strategy for reaching out to relevant stakeholders, including an assessment of the budget, resources and costs associated with Smart City development.
	• There is a designated senior manager or management team responsible for implementing the Smart City strategy, coordinating and overseeing all Smart City initiatives and ensuring synergy between them.
	• A common terminology related to Smart City and a common reference model were agreed upon.
	• Priorities for Smart City development in terms of priority areas, technologies and initiatives have been identified.
	• There are ready-made assessment plans and KPI targets for each maturity level for Smart City development.
	• The KPI values of the city's current performance are collected and recorded as baseline results.
Level 2	The goal that the city must achieve at this level is to align Smart City initiatives with the city's strategy, such as deploying ICT infrastructure to support Smart City operations and development activities.
	Achievements at this level may include the following issues:
	• An infrastructure development plan is made ready in accordance with the city's Smart City strategy.
	• Key ICT infrastructure is identified to support Smart City initiatives.
	• ICT infrastructures are capable of operating independently to provide various Smart City services.
	• ICT infrastructure registries are built and periodically updated.
	• Self-assessment of ICT infrastructure and services is conducted periodically.
	• Improved performance on maturity level 2 KPI targets has been achieved in accordance with the plan in the city's Smart City strategy.
Level 3	The goal that the city must achieve at this level is to implement specific Smart City initiatives, providing Smart City services based on ICT infrastructure through, for example, local social service centers, mobile applications and web portals.
	Achievements at this level may include the following issues:
	• City departments or specific authorized organizations, as well as private sector companies, are building separate platforms or systems to systematically manage resources and data.
	• Services are made available through various channels, such as mobile apps, web portals, service platforms and community terminals.

(*Continued*)

Table 3.4 (Continued)

Maturity level	Characteristics
	• Services are being upgraded by enabling improved functionality.
	• Application performance is monitored and analyzed to improve performance and service quality.
	• User satisfaction assessments are periodically conducted for target communities.
	• Performance improvements have been achieved against the KPI targets for maturity level 3 as planned in the city's Smart City strategy.
Level 4	The goal for the city at this level is to ensure that systems and data are integrated to provide city services. Technologies such as the Internet of Things (IoT), cloud computing, artificial intelligence and other advanced technologies can be used to improve service quality and interoperability.
	Achievements at this level may include the following issues:
	• ICT infrastructure interoperability has been achieved.
	• Collaboration between infrastructures, systems and/or communities has been established.
	• A cross-domain platform and applications were provided.
	• Open data are available to the public from a variety of sources as needed.
	• Stakeholder and service provider satisfaction assessments are conducted periodically.
	• Performance improvements have been achieved against the KPI targets for maturity level 4 in accordance with the plan in the city's SSC strategy.
Level 5	The objective the city must achieve at this level is to continuously improve Smart City operations. Each of the city's services is being studied to identify ways to increase value for citizens while reducing operating costs. Collaboration between systems, data, innovative services and applications is expected to increase the value of the city and improve citizen satisfaction and quality of life. The efficiency and effectiveness of city management are improved to further contribute to the city's long-term Smart City vision.
	Achievements at this level may include the following issues:
	• Services, applications and collaboration based on collaborative systems are constantly being improved.
	• Management and action based on qualitative and quantitative analysis are effectively established.
	• Continuous improvement of services and applications is made possible by technology.
	• A systematic evaluation process has been established to carry out continuous improvement and performance evaluation.
	• The results of the assessments and evaluations are analyzed, and appropriate action plans are implemented as part of the city's Smart City strategy.

Source: Smart (2019).

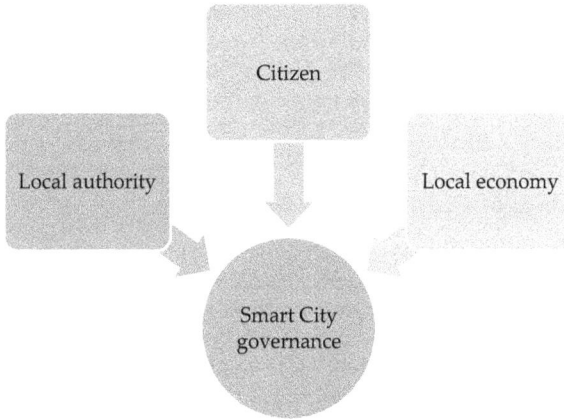

Figure 3.1 Areas of smart city management.

Table 3.5 Dimensions of key areas of smart city management

Area	Dimensions
Local authority	• Cost reduction • Improved government transparency • Increased collaboration • Improved decision-making • Disseminating knowledge and expertise • Improved work efficiency
Citizen	• Flexibility • Social cohesion • Lifelong learning opportunities • Improved health conditions and independence • Better community connectivity • Increased employment opportunities
Local economy	• Promote innovation • Catalyze development of new products and services • Engage and leverage SME community • Accelerate new business startups
Local authority and citizen	• Participation in public life • Resilient public services • Social equity
Local authority and local economy	• Leveraging private funding • Inward investment
Citizen and local economy	• Increased economic activity
Local authority and citizen and local economy	• Improved resource efficiency • Sustainable mobility • Environmental sustainability • Economic prosperity

Source: Overview (2016).

Yoo, 2021). Figure 3.1 shows the three most important areas that make up the proper management of a smart city. These include local authority, citizen and local economy. Table 3.5 shows the aspects that make up each area and those that are involved in more than one of these three areas.

3.2 The Smart City concept in the strategies of the cities of Central and Eastern Europe

In Poland today, a number of cities have developed their Smart Cities strategies, outlining the concept of implementing the concept within a given city. This chapter will present several examples on a case study basis of Smart City strategies developed by cities in Central and Eastern Europe. Special attention will be paid to the examples of Polish cities.

Most cities in Central and Eastern Europe do not have a separate Smart City development strategy. Rather, they only consider certain aspects of the Smart City concept that are included in the overall city development strategy. Only a few cities have a special strategy dedicated to Smart City development. In this section, we will focus on such documents.

An example of a comprehensive Smart City strategy developed by a Polish city is the Smart City Strategy of the City of Kołobrzeg. It was released in 2019 and the document is 184 pages long. The City of Kołobrzeg has adopted the following Smart City vision: *Kołobrzeg is a city, people and technology constituting one organism, a vibrant system of connections and dependencies. Kołobrzeg is a city co-managed by its residents, open to the needs of residents and tourists, reaching for intelligent solutions to support its development. It is also an anticipatory city, willing to anticipate the needs of its current and future residents* (Smart City Strategy of the City of Kołobrzeg, 2019). For the strategy presented, the city has established a number of strategic goals, which are presented in Table 3.6, along with Smart City challenges in this regard.

Another example of a Polish city that has a strategy dedicated to Smart Cities solutions is Kielce. In 2022, the Kielce City Council adopted a document called 'Kielce City Development Strategy 2030+ towards Smart City'. Table 3.7 shows the strategic goals adopted in the Kielce city strategy, along with the most important Smart City activities the city intends to undertake.

It is interesting to note that having a Smart City strategy does not always depend on the size of the city or its economic importance. For example, Poland's capital Warsaw does not have a separate Smart City strategy, and the city's overall strategy 'Warsaw 2030 – Strategy' only presents a definition of a smart city, while in its content, no points directly refer to the Smart City concept (Warsaw 2030 Strategy, 2018). This does not mean that Warsaw is not implementing numerous Smart City projects

Table 3.6 Strategic objectives of the Smart City Strategy of the City of Kołobrzeg

Strategic objective	*Challenges for Smart City development*
Diversified and innovative economy of the future	• Building smart economic specialization related to active support for selected sectors of the economy with the greatest potential for innovation. • Investing in the development of services of the future (ICT industry, IT, creative industries). • Creating incubators of local culture and handicrafts and increasing the role of culture and art in Kołobrzeg's economy. • Promoting Kołobrzeg as a place with exceptional service quality. • Creating space for the development of initiatives, events and fairs related to wellness, health and personal development. • Strengthening economic networks, aimed at forming value-added, innovative services and products. • Creating attractive living conditions for engineering and arts graduates through excellent living conditions and attracting investors.
The best conditions for living and relaxing on the coast	• Sustainable use of the city's space and resources, supported by a transparent information policy. • Developing a sustainable and low-carbon mobility system, based on public transportation of the city, region and country, a developed bicycle transportation system and management systems. • Shaping the city's cultural policy in cooperation with local art communities and other entities. • Improving the safety of residents and tourists.
Kołobrzeg managed by residents	• Active, experienced and cooperating with each other and the public sphere cadres of local NGOs and leaders, supported by infrastructural and integration solutions. • The universality of residents' participation in the city's development processes and activities. • Competent public services, skillfully using innovative tools and technologies in a participatory model of city management.

Source: Smart City Strategy of the City of Kołobrzeg (2019).

and activities. In recent years, for example, solutions, such as 'Warsaw Towards a Smart City (2018)', have been introduced in this area:

• Introduction of the Veturilo system – It is one of the largest urban bicycle systems in Europe. It is an important part of Warsaw's transportation ecosystem. Thanks to the expansion of bicycle paths, whose network counts more than 500 km in Warsaw, the system allows

Table 3.7 Strategic objectives of the Smart City Strategy of the City of Kielce

Strategic objective	Selected Smart Cities activities
A city for everyone	• Promoting tolerant attitudes regarding freedom of belief, differing attitudes and civil liberties. • Adapting city infrastructure to accommodate strollers and introducing facilities for people with disabilities. • Increasing the external institutional efficiency of Kielce City Hall (improving the quality of public service delivery). • Supporting the area of residential and labor migration in Kielce by providing potential migrants with access to reliable information on the opportunities awaiting them. • Eliminating the digital exclusion of Kielce residents caused by lack of knowledge and skills in using modern technologies. • Accelerating the construction of new bicycle routes that enable the use of the bicycle as a full-fledged and safe means of transportation in every settlement of the city. • Improving the network of public transport system connections from individual neighborhoods to downtown. • Spreading sharing economy solutions in the availability of individual transportation. • Eliminating disparities in the level of infrastructural development of individual settlements (especially the so-called peripheral ones) in terms of basic services such as water supply, heating networks, sewerage and the Internet. • Improving the quality of life in large slab buildings by designing elements such as neighborhood spaces, semi-public and public spaces, and nodal points. • Stimulating social activity among residents through promotional activities and more information to the public about the activities undertaken by the City Hall. • Developing and promoting tools for resident participation in decisions on city development both on site and remotely. • Increasing awareness of Kielce residents of the structure of the City Hall's operations and how to get in touch.
An efficient and creative city	• Increasing the internal institutional efficiency of Kielce City Hall (revision and development of procedures, development of an innovative culture, combating siloed activities, effective use of data). • Using the full potential of the data held by the office to make decisions supported by reliable, up-to-date and comparable data and to better predict and prevent future problems. • Optimizing the tasks and structure of the office to improve competence and reward. • Reviewing needs for possible ICT investments. • Enabling, stimulating and encouraging cooperation of stakeholders from the world of business, science, and business environment institutions and social organizations with the Kielce City Hall.

(*Continued*)

Table 3.7 (Continued)

Strategic objective	Selected Smart Cities activities
	• Supporting Kielce's innovation generation and implementation ecosystem by accelerating and incubating Kielce's SMEs.
	• Developing and promoting tools for the participation of residents and housing cooperatives in city development decisions.
	• Implementing intelligent transportation system solutions, road safety and public transportation integration.
	• Integrating processes and offerings to support innovation activities.
	• Digitizing services offered by the City Hall, in particular providing easy-to-use solutions for the elderly (e.g., teleconsultations and video meetings).
	• Conducting activities to integrate Kielce residents using digital tools.
	• Expanding the active (equipment) and passive (teletechnical network) ICT infrastructure of Kielce City Hall and organizational units of Kielce Municipality.
	• Improving the processes of how residents handle matters at the office by digitizing the administrative service process and making e-services available at the highest possible level of maturity.
	• Increasing the share of energy obtained from RES (e.g., photovoltaic or biofuel).
	• Combating air pollution in Kielce by stepping up efforts against smoking with prohibited materials.
	• Upgrading the city's lighting systems toward energy efficiency.
A city on the rise	• Addressing the effects of climate change due to global warming by eliminating points of excessive heat and, for example, adapting infrastructure to extreme weather conditions including the development of green and blue infrastructure.
	• Increasing the share of energy obtained from RES (e.g., photovoltaic or biofuel).
	• Upgrading the city's lighting systems toward energy efficiency.
	• Maintaining biodiversity throughout the city.
	• Increasing the share of energy obtained from RES (e.g., photovoltaic or biofuel).
	• Combating air pollution in Kielce by stepping up efforts against smoking with prohibited materials.
	• Upgrading the city's lighting systems toward energy efficiency.
	• Maintaining biodiversity throughout the city.
	• Implementing and promoting the concept of the Closed Circuit Economy.

(*Continued*)

Table 3.7 (Continued)

Strategic objective	Selected Smart Cities activities
	• Planting trees in the streets. • Promoting water retention, including small retention, measures to stop rainwater runoff. • Increasing the number of Selective Waste Collection Points. • Improving air quality through, among other things, replacement of central heating networks and thermal modernization of residential and public use buildings. • Adapting transportation infrastructure to the requirements of tourists (e.g., directional signs leading to tourist attractions and maps to facilitate movement on public transportation).
A city where you want to live	• Improving the enforcement of fines for littering in urban areas. • Transforming the city into a cohesive and urban and architecturally attractive area. • Introducing the ability to simply and quickly report neglected areas of the city that require immediate attention from the City Hall. • Conducting regular revitalization activities, especially for historic buildings. • Developing a system for managing advertising in public places – system implemented under a participatory advertising resolution and on the basis of regular consultations with residents. • Increasing the convenience of public transportation travel through investment in zero- and low-emission rolling stock, infrastructure, modern equipment and IT solutions for public transportation. • Integrating public transportation modes in a way that allows travel with 'transfers' without unnecessary delays in the form of waiting for the next trip, especially in the morning hours. • Increasing the safety of pedestrians at crossings occurring on busy roads by illuminating them and providing additional traffic lights to increase their visibility to drivers. • Evaluating current Mobility Plan solutions.

Source: Kielce City Development Strategy 2030+ towards Smart City (2022).

efficient access to various parts of the city, providing an alternative to means of public transport.
• Mobile applications – Warsaw has a number of mobile applications that make it easier for public transport passengers to buy a ticket, pay for city parking, check departure times, plan a route or estimate the actual arrival time of a bus or streetcar based on vehicle location data provided by Warsaw.

- Intelligent District Heating Network – It is a joint investment by Warsaw and Veolia Energia Warszawa S.A., completed in autumn 2017. By modernizing the existing network, installing appropriate equipment and using infrastructure management applications, it is possible to optimize the use of resources and reduce carbon dioxide emissions in Warsaw by at least 14,500 tons per year, equivalent to planting 1 million trees.
- Warsaw 19115 City Contact Centre – It is a state-of-the-art, multi-channel contact center that enables people to contact the city 24/7 via phone, email, chat and mobile application. The latter is a convenient tool for signaling city faults, suggesting tree planting sites or reviewing project proposals submitted to the participatory budget. In 2017, the 19115 system was recognized by the German editorial board of the industry magazine *CRN* as one of the five most innovative digital transformation projects in Europe.
- Warsaw participatory budget – It is a tool that has been developed for several years to support community involvement in shaping the development of Warsaw. It allows citizens to be involved in the process of co-determining the city's spending. Warsaw participatory budget since 2015 is based on an application that allows residents to submit initiatives and desired actions in the city.
- Otwarte dane po warszawsku (Open data the Warsaw way) – Since 2015, Warsaw has made over 200 datasets available through a dedicated online platform. This allows anyone to get quick access to data from official sources, including transportation, education, history, culture, entertainment, real estate and community projects.

Among Slovak cities, an example of a city with Smart City strategies is the city of Trenčín. The relevant document was created in 2016 and is called Stratégia implementácie SMART technologii v Meste Trenčín (2016). This strategy, as a basic medium-term program document for the implementation of smart technologies in the city of Trenčín, takes into account the strategic documents of the Slovak Republic. At the same time, as one of the regional development documents processed at the local level, it is based on documents processed at higher hierarchical levels.

The area of education and coordination of the implementation of Smart City concept within the city of Trenčín is currently the least developed area in the sphere of IoT solutions. However, the city lacks an entity that would integrate individual stakeholders and that would be directly responsible for the implementation of Smart City solutions in the city's daily life. Currently, IoT solutions are only partially implemented. In some cases, each entity operating in the city is implementing its own projects, which are often not compatible with each other. On the other hand, in the recent period, the first results of the city government's work can be seen, which is trying to coordinate the whole area, first at the strategic level and later at lower levels. The city's strategic goals are shown in Table 3.8.

The analyzed cities have carried out numerous projects in the Smart Cities area in recent years. For example, the city of Trenčín implemented the SMART Street project in cooperation with Slovak Telekom, which enabled the creation of a smart street in Trenčín (Palackého Street), where several smart technologies were tested. The effect of this pilot was primarily that the two cooperating entities gained a lot of experience and practical knowledge in implementing smart technologies in public administration. The disadvantage of the project, however, was that both the technologies and smart data are not owned by the city. Therefore, the city plans to apply the principle of municipal ownership of smart data in other future projects (if possible).

Another example of the use of Smart City solutions in Trenčín is lighting issues. During the reconstruction of the square, the city invested in

Table 3.8 Strategic objectives of the Smart City Strategy of the City of Trenčín

Strategic objective	Characteristics of the objective in relation to Smart City
Basic technical infrastructure	Modernized basic technical infrastructure with the use of smart technologies to increase the efficiency of its operation leading to improved living standards for city residents.
City energy management	Energy management in the city using smart technologies that will enable highly efficient operation of energy facilities.
Social infrastructure	High-quality social infrastructure using smart technologies that will provide higher efficiency in service delivery and better living conditions for city residents.
Public policy management	Introducing modern technologies into the process of conducting public administration activities, the decision-making process and public administration's contact with the society in order to raise the public standard of living.
Environmental policy of the city	The environmental policy of the city of Trenčín is based on the elimination of negative impacts affecting the quality of the environment and the level of health and quality of life of its residents through the use of smart technologies.
Coordination and education on the Smart City concept	Creating a methodological, integration and communication platform that will enable the gradual introduction of new technologies and processes into city life with conceptual education on smart technologies within the city.

Source: Stratégia implementácie SMART technológií v Meste Trenčín (2016).

smart lighting columns. With these columns, the functional characteristics of the lighting are adjusted according to smart rules not only to ensure maximum efficiency but also to ensure the safety of residents through smart 'reactions', such as lighting up in case of security incidents in the square. At the same time, the lighting columns are a source of Wi-Fi Internet signal and contain cameras, speakers and an electric car charger. The columns are also modular, so they can be customized. After the pilot project was completed, the municipality proceeded to install smart lighting columns in other parts of the city, such as the Municipal Atrium. It also plans to implement them in the near future when redeveloping the surroundings of the city center (Stratégia implementácie SMART technológií v Meste Trenčín, 2016).

Another example of a Slovak city with a Smart City strategy is Dunajská Streda. The strategy was developed in 2020 and is titled 'Stratégia zavádzania SMART technologii v meste Dunajská Streda (2020)'. The city has also been involved in numerous Smart City projects. Worth mentioning here are solutions such as 'Stratégia zavádzania SMART technológií v meste Dunajská Streda (2020)':

- City mobile application – The application contains news about the city, events and recreational opportunities. You can quickly and easily find up-to-date information about local government institutions and services with its help. For example, you can find restaurants, cafés or office opening hours.
- Electronization of services in the city of Dunajská Streda – The project was to be part of the informatization of society and the construction of eGovernment in Slovakia. Its implementation was to ensure the availability of electronic services of the city of Dunajská Streda and their greater usability. The intention of the project was to computerize part of municipal services. Upon completion of project activities, there will be a reduction in the administrative burden on citizens, businesses and the municipality, including employees performing specific tasks, streamlining and improving the management of the agenda and provision of eGov services, simplifying the way in which new electronic municipal services are created and provided, speeding up the processing of requests and handling of services, and minimizing the provision of duplicate data entered by users.

In the case of Hungary, the capital Budapest has a document titled 'Smart Budapest Okos Város Keretstratégia (2021)'. Table 3.9 includes selected goals of Budapest's Smart City strategy.

In the case of the Czech Republic, an example of a city with a Smart City strategy is the city of Klatovy. The strategy was created in 2021 and is titled 'Strategie Smart City Implementační část Města Klatovy (2021)'.

Table 3.9 Strategic objectives of the Smart City Strategy of the City of Budapest

Strategic objective	Characteristics of the objective in relation to Smart City
Innovation	Provide Budapest with a sustainable, durable and diverse system of jobs in innovative industries.
	A city that is internationally competitive. The public sector plays a coordinating role in supporting and implementing innovation in the city.
	Companies should be prepared for technological challenges through access to information and knowledge.
Sustainable development	Reduce resource utilization by avoiding waste of resources by minimizing energy waste.
	Increase the share of renewable energy sources in the energy mix.
	Increase the participation of urban entities in the re-education of greenhouse gas emissions.
Transportation system	Use smart applications to manage transportation in the city.
	Reduce noise pollution through better transportation planning and land use.
Urban zone planning	Achieve an affordable housing system.
	Good planning of public spaces – pedestrian-friendly zones, green areas, etc.

Source: Smart Budapest Okos Város Keretstratégia (2021).

The strategy includes the following thematic areas from a Smart City perspective (Strategie Smart City Města Klatovy Analytická část, 2021):

- city administration, management and development (including organizations established or owned by the city);
- administration, management and development of administrative bodies (including organizations established or owned by the city);
- business support and cooperation with entrepreneurs;
- city management and financial planning (including organizations established or owned by the city);
- communication between the city and the office;
- citizen participation;
- ICT solutions and data work for the city and the authority.

The Smart City strategy for the city of Klatovy devotes a special place to the issue of electronization of city services. For example, the possibility of electronic ordering of the following selected documents has been introduced: travel documents, driver's licenses and identity cards. Electronic versions are also available, as well as the possibility to fill out selected forms electronically. An electronic attendance system has been introduced in the office. The system introduced in the city also includes the processing of electronic vacation requests, including electronic signatures.

In the case of the Czech Republic, the capital Prague also has a Smart City strategy – the document is called Smart Prague 2030 – and includes a concept for implementing Smart City projects in the city. The document is based on the following six areas that are favored in Smart Prague – Mobility of the Future (Smart Prague Action Plan 2030, 2019):

1 mobility of the future
2 smart energy and buildings;
3 waste-free city;
4 attractive tourism;
5 people and the urban environment;
6 data area

Of these six areas, the data area is very important because it provides connectivity in all areas. The implementation of Smart Cities is mainly provided by IT solutions.

In determining the relevant targets and indicators, the actual state of affairs in each area was based on the data contained in the Smart Prague Index and the information provided by the individual management

Table 3.10 Strategic objectives of the Smart City Strategy of the City of Uherský Brod and measures of their achievements

Strategic objective	Sample measuring factors
Development of the city's built-up area	• Traffic volume in selected locations • Waiting time at the intersection • Amount of emissions produced by automobile traffic • Number of available parking spaces • Number of safe pedestrian crossings • Share of buildings with untapped energy savings potential
Community life in the city	• Number of visits to the city's website • Increased use of IT tools by seniors in smart healthcare (e.g. e-prescription and SOS app) • Number of service e-portal registrations by citizens and their active use • Club occupancy (for children/seniors/mothers with children) • Number of startup companies established • Modernization of schools and school facilities
Quality of Life	• Participation in cultural and sports events • Scope of use of alternative energy sources • Number of roofs of municipal buildings with photovoltaics • Number of implementations of anti-drought measures and their functionality

Source: On basis: Strategie Smart city města Uherský Brod. (2019), Overview (2016).

organizations. With each indicator, a guarantor is established, i.e., the organization that manages the indicator, and is therefore responsible for meeting it. As a starting point, the values of 2017, 2018 and 2019 are used, based on the information contained in the Smart Prague Index, with targets set for what level of values the Smart Prague area should reach by 2030. These targets were consulted with the relevant management organizations and represent an expert, preliminary assessment of the target state in 2030. With the targets thus set, progress can be tracked annually as part of the implementation of the 2030 action plan. The set target can thus be monitored annually (Smart Prague Concept by 2030, 2019).

Another example of a Czech city with a Smart City strategy is Uherský Brod. The strategy was created in 2019 (Strategie Smart city města Uherský Brod., 2019). Table 3.10 shows the city's strategic goals and sample measures of their achievement related to the implementation of the Smart City concept.

Bibliography

Albino, V., Berardi, U., Dangelico, R.M. (2015). Smart cities: Definitions, dimensions, performance, and initiatives. *Journal of Urban Technology, 22*(1), 54–73.

Arora, V. (2021). *What is smart city governance?* https://www.planetcrust.com/what-is-smart-city-governance?utm_campaign=blog, [access data: 23.10.2022].

Bokhari, S.A.A., Myeong, S. (2022). Artificial intelligence-based technological-oriented knowledge management, innovation, and e-service delivery in smart cities: Moderating role of e-governance. *Applied Sciences, 12*(17), 8732

Caragliu, A., Del Bo, Ch., Nijkamp, P. (2011). Smart cities in Europe. *Journal of Urban Technology, 18*(2), 15–38.

Carrato-Gómez, A., Roig-Segovia, E. (2022). From the sustainable city to the hub city: Obsolescence and renewal of urban indicators. *Ciudad y Territorio Estudios Territoriales, 54*(213), 563–578.

Clark, R.Y. (2017). Measuring success in the development of smart and sustainable cities. In: Cronin, M.J., Dearing, T.C. (eds) *Managing for Social Impact. Innovations in Responsible Enterprise.* Springer International Publishing: Cham, 152–178.

Dameri, R.P. (2013). Searching for smart city definition: A comprehensive proposal. *International Journal of Computers & Technology, 11*(5), 75–128.

Ependi, U., Rochim, A.F., Wibowo, A. (2022). *Smart City Assessment for Sustainable City Development on Smart Governance: A Systematic Literature Review,* 2022 International Conference on Decision Aid Sciences and Applications, DASA 2022, 1088–1097.

Faraji, S.J., Jafari Nozar, M., Arash, M. (2021). The analysis of smart governance scenarios of the urban culture in multicultural cities based on two concepts of "cultural intelligence" and "smart governance". *GeoJournal, 86*(1), 357–377.

Fonseca, D., Sanchez-Sepulveda, M., Necchi, S., Peña, E. (2021). Towards smart city governance. Case study: Improving the interpretation of quantitative traffic measurement data through citizen participation. *Sensors, 21*(16), 5321.

Founoun, A., Hayar, A., Essefar, K., Haqiq, A. (2022). *Agile Governance Supported by the Frugal Smart City*, Lecture Notes in Networks and Systems, *334*, 95–105.

Hajduk, S. (2016). Selected aspects of measuring performance of smart cities in spatial management. *Business and Management, 2*, 8–16.

Hajduk, S. (2020). Modele smart city a zarządzanie przestrzenne miast. *The Polish Journal of Economics, 302*(2), 123–139.

He, W., Li, W., Deng, P. (2022). Legal governance in the smart cities of China: Functions, problems, and solutions. *Sustainability, 14*(15), 9738.

Hollands, R.G. (2008). Will the real smart city please stand up? Intelligent, progressive or entrepreneurial? *City, 12*(3), 128–136.

Komninos, N. (2014). *The Age of Intelligent Cities: Smart Environments and Innovation-for All Strategies.* Routledge: London.

Koncepce Smart Prague do roku 2030. (2019). https://smartprague.eu/files/koncepce_smartprague.pdf, https://www.smartprague.eu/action-plan, [access data: 23.10.2022].

Laurini, R. (2021). A primer of knowledge management for smart city governance. *Land Use Policy, 111*, 104832.

Lim, S.B., Yigitcanlar, T. (2022). Participatory governance of smart cities: Insights from e-participation of Putrajaya and Petaling Jaya, Malaysia. *Smart Cities, 5*(1), 1–89.

Maurya, K.K., Biswas, A. (2021). Performance assessment of governance in Indian smart city development. *Smart and Sustainable Built Environment, 10*(4), 653–680.

Mohsin, B.S., Ali, H., Al Kaabi, R. (2019). *Smart city: A review of maturity models*, 2nd Smart Cities Symposium (SCS 2019). https://doi.org/10.1049/cp.2019.0209

Nina, X., Hao, Z., Huije, L., Rongxial, Y., Jia, W., Zhongke, F. (2022). Performance analysis of smart city governance: Dynamic impact of Beijing 12345 hotline on urban public problems. *Sustainability, 14*(16), 9986.

Overview of the Smart Cities Maturity Model. (2016). Urban Tide, https://static1.squarespace.com/static/5527ba84e4b09a3d0e89e14d/t/55aebffce4b0f8960472ef49/1437515772651/UT_Smart_Model_FINAL.pdf, [access data: 23.10.2022].

Saadah, M. (2021). Artificial intelligence for smart governance; towards Jambi Smart City. *IOP Conference Series: Earth and Environmental Science, 717*(1), 012030.

Shi, D., Cao, X. (2022). Research on the effectiveness of government governance in the context of smart cities. *Proceedings of SPIE –The International Society for Optical Engineering, 12165*, 121651F.

Smart Budapest Okos Város Keretstratégia. (2021). https://budapest.hu/Documents/V%C3%A1ros%C3%A9p%C3%ADt%C3%A9si%20F%C5%91loszt%C3%A1ly/Smart%20Budapest%20Keretstrat%C3%A9gia%202019.pdf, [access data: 23.10.2022].

Smart Prague Action Plan 2030. (2019). https://www.smartprague.eu/action-plan, [access data: 23.10.2022].

Smart sustainable cities maturity model. (2019). International Telecommunication Union, file:///C:/Users/radek/Downloads/T-REC-Y.4904–201912-I!!PDF-E.pdf, [access data: 23.10.2022].

Stratégia implementácie SMART technológií v Meste Trenčín. (2016). https:// trencin.sk/wp-content/uploads/2021/01/SMART_strategia_Trencin-2021-01-07-update.pdf, [access data: 23.10.2022].

Strategia Rozwoju Miasta Kielce 2030+ w kierunku Smart City. (2022). https:// www.kielce.eu/pl/dla-mieszkanca/samorzad/dokumenty-strategiczne-i-operacyjne/dokumenty-strategiczne/strategia-rozwoju-miasta-kielce-2030.html, [access data: 23.10.2022].

Strategia Smart City Miasta Kołobrzeg. (2019). http://umkolobrzeg.esp.parseta. pl/uploads/media/Strategia_Smart_City_Miasta_Kolobrzeg_-_projekt_dokumentu.pdf, [access data: 23.10.2022].

Stratégia zavádzania SMART technológií v meste Dunajská Streda. (2020). https://dunstreda.sk/files/strategia_zavadzania_smart_technologii_ds.pdf, [access data: 23.10.2022].

Strategie Smart City Implementační část Města Klatovy. (2021). https://www. klatovy.cz/mukt/user/strategie/sc/Klatovy-Smart_City-implementacni_cast_ ke_schvaleni.pdf, [access data: 23.10.2022].

Strategie Smart City Města Klatovy Analytická část. (2021). https://www.klatovy. cz/mukt/user/strategie/sc/Klatovy-Smart_City-analyticka_cast.pdf, [access data: 23.10.2022].

Strategie Smart city města Uherský Brod. (2019). https://www.ub.cz/Public/ docs/KOMPAS/Strategie_Smart_City.pdf, [access data: 23.10.2022].

Vitálišová, K., Sýkorová, K., Koróny, S., Rojíková, D. (2022). Benefits and obstacles of smart governance in cities, lecture notes of the institute for computer sciences. *Social-Informatics and Telecommunications Engineering, 442,* 366–380.

Vujković, P., Ravšelj, D., Umek, L., Aristovnik, A. (2022). Bibliometric analysis of smart public governance research: Smart city and smart government in comparative perspective. *Social Sciences, 11*(7), 293.

Waarts, S. (2016). *Smart city development maturity,* http://arno.uvt.nl/show. cgi?fid=143408, [access data: 23.10.2022].

Warszawa 2030 Strategia. (2018). https://um.warszawa.pl/documents/ 56602/38746844/Strategia+%23Warszawa2030.pdf/990ffe5b-835b-515b-d899-0a34d600f591?t=1646311753449, [access data: 23.10.2022].

Warszawa w kierunku Smart City. (2018). https://pawilonzodiak.pl/wp-content/ uploads/2018/10/kfsmartwawa2018pl.pdf, [access data: 23.10.2022].

Willis, K.S., Nold, C. (2022). SmartAirQ: A big data governance framework for urban air quality management in smart cities. *Frontiers in Environmental Science, 10,* 785129.

Yoo, Y. (2021). Toward sustainable governance: Strategic analysis of the smart city Seoul portal in Korea. *Sustainability, 13*(11), 5886.

4 Characteristic of Central and Eastern Europe cities as research sites

4.1 Determinants of smart cities development in Central and Eastern Europe

This section will present population statistics for selected countries in Central and Eastern Europe. The conditions for the development of the Smart City concept in these countries will also be presented in a synthetic way. The following countries included in Central and Eastern Europe were selected for the analysis: Poland (in the next section), Czech Republic, Slovakia, Hungary and Lithuania.

In 2022, the Czech Republic has 27 cities, 582 towns and 231 market towns. The Czech Republic's ten largest cities are listed in Table 4.1.

In the Czech Republic, the development of the Smart City concept is being implemented through the Czech Smart City Cluster (CSCC). The Cluster (CSCC) is tasked with developing a unique partnership between companies, state administration, local government, knowledge institutions and city residents (Czech Smart City, 2022). The Cluster promotes the idea of Smart City in the Czech Republic. It seeks to build smart cities where infrastructure and social and technological solutions make people's lives easier and support sustainable economic growth (Pacovský and Jolič, 2021). These trends should improve the quality of life in cities for all residents and cities, thus becoming a pleasant environment to live and work.

Cluster members focus on the integration of smart technologies, such as in the areas of energy, smart buildings, transportation and ICT. As part of their projects, they are transforming traditional, isolated infrastructures into highly integrated systems that intervene at all levels, from buildings and technology units to municipalities to the regional and then state levels.

A number of Smart City projects are being implemented in the Czech Republic. Table 4.2 lists examples of Smart City projects implemented in Prague.

There are 141 cities in Slovakia. As with the Czech Republic, Table 4.3 lists population figures for Slovakia's ten largest cities in 2022.

DOI: 10.4324/9781003358190-4

Table 4.1 The largest cities in the Czech Republic

City	Number of inhabitants
Prague	1,275,406
Bron	379,466
Ostrava	279,791
Plzeň	168,733
Liberec	102,951
Olumunc	99,496
České Budějovice	93,426
Hradec Králové	90,596
Ústí nad Labem	90,378
Pardubice	88,520

Source: Population (2022).

Table 4.2 Selected Smart City projects implemented in Prague

Project	Description
Complex energy management in buildings	The project will be implemented in several stages. First, a review of energy and water consumption in a selected sample of buildings will be carried out. In the next phase, potential efficiency measures will be identified and proposals made for their implementation. Responsible city officials will then be introduced to the measures to be implemented. This is the most appropriate way to implement such projects. Benefits: • more efficient operation and management of the city of Prague's buildings, • optimization of energy and water consumption, • general development of energy management, • cost savings.
Digital energy consumption measurements	The project will provide a new multipurpose metering system for selected buildings in Prague. The system will provide complete and continuous measurements of the energy consumed, thus creating the conditions for flawless monitoring of all energy consumption in a given building. Customers will be able to monitor their consumption and related activities at all times on the free-of-charge PREměření web portal. Benefits: • effective energy readings, • reduced energy consumption, • effective management of the urban real estate, • estimated savings: 15%.
Smart waste collection	Waste collection routes will be optimized using a special app. With sensors placed directly in waste containers, which will be connected online to the app, the app can intelligently plan the collection route.

(*Continued*)

Table 4.2 (Continued)

Project	Description
	Benefits: • reduced cost of waste collection, • reduced environmental impact, • improved road accessibility, • data collected for further processing and optimization of public spaces.
Prague Visitor Pass	The Prague Visitor Pass is a multifunctional card that will be physical and electronic and will be available in several time and age variants. Tourists will be able to use it as an entrance ticket to places of tourist importance (museums, galleries, zoos, etc.), as a CTS ticket within the capital city of Prague, and apply for discounts at selected intermediaries (circuitous bus or boat tours, guide services and others). The PVP system includes web and mobile applications, with a full-fledged online store for both platforms. The mobile app can also be used as the system's ID and thus replace the plastic PVP card. Benefits: • The implementation of the Prague Visitor Pass will lead to an improvement in the efficiency of services in the area of tourism in the capital city of Prague; its sales should increase awareness of the location, tourist attractions and services. • With higher potential primarily in terms of user comfort and the service portfolio extensions offered, greater popularity and, thus, sales are expected.
Intelligent traffic analysis	The project aims to solve the problem of acquiring statistical data on the traffic of Prague's transport modes with a high-quality tool that will be able to comprehensively determine the most important traffic data and thus provide support for responsible decision-making on traffic issues. The tool must meet the basic criterion for success, which is reliability. It is necessary to track traffic data 24/7 in real time to provide an objective and high-quality picture of traffic in Prague. The tool will allow for the acquisition of traffic data for securing high-quality decision-making in terms of: • road infrastructure adjustments, • landscape planning, • traffic modeling, • increased traffic flow and safety, • crisis management, • traffic development and further tracking of the entire transportation system of the capital city of Prague, • reduced stress of traffic-related emission.

(*Continued*)

Table 4.2 (Continued)

Project	Description
Autonomous mobility in the capital city of Prague	The first phase of the project will create materials for city officials making decisions on technological support in the perspective of moving automated vehicles. The creation of an above-ground transportation test area for autonomous vehicles is an important step toward autonomous mobility. The pilot project will be used to test vehicles equipped with an autonomous driving system in real urban traffic. A team of experts on autonomous mobility will be established to deal with the province's organization and regulation. Benefits: • direct support of autonomous vehicles in urban traffic, • the environment created for the official implementation of tests of autonomous vehicles in real traffic, • mobility trends recognized in the development of the Prague capital city area, • valuable lessons gathered for future collection of road infrastructure upgrades, • increased Prague's competitive strength compared to other European cities.

Source: Smart Prague (2022), Rohlena and Frkova (2014), Peterka (2019), Action plan (2020), Svítek et al. (2020), Štěpánek (2020), Vácha and Kandusová (2018).

Table 4.3 Slovakia's largest cities

City	Number of inhabitants
Bratislava	423,737
Košice	236,563
Prešov	94,718
Nitra	86,329
Žilina	85,985
Banská Bystrica	82,336
Trnava	69,785
Trenčín	58,278
Martin	54,618
Poprad	57,431

Source: Population of Cities in Slovakia (2022).

Table 4.4 lists selected Smart City projects implemented in Bratislava. Bratislava's vision of a smart city is based on the concept of 'the city is a living system' in which the environment and natural resources, the city's development (urbanization) and infrastructure, residents, communities and society as a whole are very closely interconnected and interact with each other. The Bratislava Smart City 2030 concept is aimed at the

Table 4.4 Selected Smart City projects implemented in Bratislava

Project	Description
Bike-Sharing System	A community program called White Bikes – a public bicycle rental system – has been established in Bratislava. The project is organized by the 'Cyklokoalicja' (Bikecoalition) Association, and its use is conditional on completing training on how to use the bikes. Currently, there are 50 white bikes and 70 registered users. In the system, White Bikes can only be rented by Bratislava residents, the third sector, non-profit organizations or anyone who works for the Bratislava municipality. The bikes can be borrowed for free. A future plan includes charging fees for returning the bikes. The goal of the bike-sharing project is for the bikes to be visible on the streets so that they can really benefit people.
SMS Parking System	Since 2010, drivers have been able to pay parking fees in marked parking lots in downtown Bratislava using their cell phones via SMS, without further registration. The parking card holder receives a return message immediately, and the price is currently set at EUR 1 per hour. Vehicles that have paid for parking through the SMS Parking System are registered in a protected database system and paid parking is inspected by authorized persons, so there is no risk of any penalty.
e-Governance	Bratislava municipality launched the e-governance project in 2015. It is an electronic city council project that contributes to saving the environment and optimizing the work of the municipality. The city's official website provides meeting materials, profiles of deputies and information about their individual vote or resolution of the City Council. The new application serves everyone – both residents and local authorities.
Smart Public Lighting	This is a concept for public street lighting with modern LED lighting that would use alternative solar energy with built-in twilight switches that turn on and off automatically, and smart control and management of public lighting. The smart lighting would regulate light intensity using motion sensors – if pedestrian paths are empty at the same time, then the light intensity will be automatically reduced to the minimum set limit. When someone appears on the road, motion sensors will detect the person and the light intensity will be increased.
Bratislava Summer University for Seniors	Free senior education. Office of Healthy City – public health and healthy lifestyle education focusing on maintaining the health of Bratislava residents, changing the lifestyle of Bratislava residents, improving the living and working environment in the city.
Fablab	Supporting an open digital technology platform for designers, developers, artists, students and the general public (individual subsidies, city budget).

Source: Golej et al. (2016a, 2016b), Bratislava (2018), Baculáková (2019, 2020).

efficient use of resources, competitiveness of the economy and increased quality of life in the city. The concept defines a framework strategy for the city – development as a Smart City – and is developed in relation to related strategic documents at the international, national and municipal levels (Bratislava, 2018).

Hungary currently has 346 cities – város in Hungarian. A list of the ten largest cities in 2022 is shown in Table 4.5. Four of the cities (Budapest, Miskolc, Győr and Pécs) have agglomerations. The Hungarian Statistical Office distinguishes 17 other areas in earlier stages of agglomeration development.

According to the city's long-term development concept, the Smart Budapest Vision will last until 2030. The point of the smart city approach is not just to implement a number of projects, but to develop an approach based on continuous adaptation, capable of responding appropriately to emerging challenges (Smart Budapest, 2017). Table 4.6 lists selected Smart City projects implemented in Budapest.

Lithuania boasts 103 cities. The term city was defined by the Parliament of Lithuania as compact urban areas with a population of more than 3,000 people, at least two-thirds of whom work in industry or the service sector. Table 4.7 provides population data for Lithuania's ten largest cities in 2022.

Vilnius has been ranked among the top 25 global cities of the future for the first time in 2021, according to fDi Intelligence, a specialized division of 'the *Financial Times*', which ranks cities based on a range of foreign direct investment indicators. After evaluating recent foreign investment attracted to the city, various economic indicators and overall business development, the Lithuanian capital ranked 24th alongside cities such as Singapore, London, Dubai, Amsterdam and New York. Vilnius has a vibrant startup community with more than 20 business hubs, gas pedals and pre-accelerators, in addition to regulatory sandboxes. Companies such as Revolut and Transfergo have established offices in Vilnius, recognizing the value of this environment (Lithuania, 2022). Table 4.8 shows examples of Smart City projects underway in Vilnius.

Table 4.5 Hungary's largest cities

City	Number of inhabitants
Budapest	1,706,851
Debrecen	199,725
Szeged	157,382
Miscolc	147,480
Pécs	138,420
Győr	132,111
Nyíregyháza	115,521
Kecskemét	108,817
Székesfehérvár	94,893
Szombathely	77,970

Source: List of localities (2022).

Table 4.6 Selected Smart City projects implemented in Budapest

Project	Description
Budapest Smart City Center of Excellence	The primary objective of the project is to prepare the establishment of a Budapest Smart City Center of Excellence, built on the innovative capacity, know-how and experience of the project partners. The new organization, SMARTPOLIS, should contribute to the creation of knowledge, knowledge transfer, as well as research, innovation and implementation projects in the CEE region in order to achieve the European goals set out in Horizon 2020.
Energy-efficient street lighting	Pilot project: • installation of 7,000 units of energy-efficient lighting fixtures, • supply of 7,000 units of energy-efficient LED street lamps, • project led by BDK, the public company responsible for street and cultural lighting.
Integrated e-mobility concept for Budapest	The Budapest municipality is committed to prioritizing environmentally friendly electric and zero-emission mobility and reducing the use of internal combustion engine cars, which are responsible for air pollution in Budapest.
LIFE HungAIRy project	Implementing the air protection plan: maintaining a proper environment and improving air quality in accordance with the human right to a healthy environment and clean air. The project's objective is to establish a database of direct public emission sources, which is the basis for a program to reduce household emissions.
Development of a climate strategy and a climate change platform	Provide a broader framework for the city's climate change adaptation and mitigation through a climate strategy. A special climate platform will be launched. The climate strategy and platform will cover all areas (i.e., transportation, energy, built environment, water and waste) to ensure sufficiently broad consultation with local and regional stakeholders and effective implementation of the city's climate change adaptation strategy.

Source: Szegvári (2017), Budapest Smart City (2015), Establishment (2015), and Turoń et al. (2022).

Table 4.7 Lithuania's largest cities

City	Number of inhabitants
Vilnius	542,366
Kaunas	374,643
Klaipėda	192,307
Šiauliai	130,587
Panevėžys	117,395
Alytus	70,747
Dainava	70,000
Eiduliai	61,700
Marijampolė	47,613
Silainiai	40,600

Source: Population of Cities in Lithuania (2022).

Table 4.8 Selected Smart City projects implemented in Vilnius

Project	Description
Smart Healthcare	Monitoring health statistics, hospital asset management, emergency notifications, monitoring vital health indicators, pharmacy inventory management and effective workforce management.
Smart Metering	Data integration between applications/devices, customized data analysis and reporting, thematic notifications, cloud storage and data exchange among multiple devices.
Retail and Logistics	Sensor-based item tracking, optimal inventory management, QoS data collection and analysis, data collection for machine learning, remote control of connected devices, smart labels and real-time fleet management.
Smart Agriculture	Sensor-based field and asset mapping, intelligent logistics and storage, remote crop monitoring and control of storage conditions.

Source: Lithuania (2022), Smart City Vilnius (2022), Dudzevičiūtė et al. (2017), and Zapolskytė et al. (2022).

4.2 Characteristic of Polish cities in the context of development determinants

As of January 1, 2022, there are 964 cities in Poland, 302 of which are independent municipalities, including 107 presidential cities, 66 cities with county rights, 37 cities over 100,000, 18 provincial cities and 11 cities that are seats of the court of appeals. In Polish cities, the executive body, which is also a public administration body, is the Mayor or City President. The decision-making and controlling body is the City Council or City and Municipal Council.

Polish cities include (Area and population, 2021):

- One city with over 1,000,000 residents: Warsaw;
- Four cities with 500,000–1,000,000 residents: Kraków, Wrocław, Łódź and Poznań;
- Six cities with 250,000–500,000 residents: Gdańsk, Szczecin, Bydgoszcz, Lublin, Białystok and Katowice;
- Twenty-six cities with 100,000–250,000 residents: Gdynia, Częstochowa, Radom, Rzeszów, Toruń, Sosnowiec, Kielce, Gliwice, Olsztyn, Zabrze, Bielsko-Biała, Bytom, Zielona Góra, Rybnik, Ruda Śląska, Opole, Tychy, Gorzów Wielkopolski, Elbląg, Płock, Dąbrowa Górnicza, Wałbrzych, Włocławek, Tarnów, Chorzów and Koszalin;
- Forty-five cities with 50,000–100,000 residents: Kalisz, Legnica, Grudziądz, Jaworzno, Słupsk, Jastrzębie-Zdrój, Nowy Sącz, Jelenia Góra, Siedlce, Mysłowice, Konin, Piła, Piotrków Trybunalski, Inowrocław, Lubin, Ostrów Wielkopolski, Suwałki, Stargard, Gniezno, Ostrowiec Świętokrzyski, Siemianowice Śląskie, Głogów,

Pabianice, Leszno, Żory, Zamość, Pruszków, Łomża, Ełk, Tomaszów Mazowiecki, Chełm, Mielec, Kędzierzyn-Koźle, Przemyśl, Stalowa Wola, Tczew, Biała Podlaska, Bełchatów, Świdnica, Będzin, Zgierz, Piekary Śląskie, Racibórz, Legionowo and Ostrołęka;

• Twenty-five cities with 35,000–50,000 residents: Świętochłowice, Wejherowo, Zawiercie, Skierniewice, Starachowice, Wodzisław Śląski, Starogard Gdański, Puławy, Tarnobrzeg, Kołobrzeg, Krosno, Radomsko, Otwock, Skarżysko-Kamienna, Ciechanów, Kutno, Sieradz, Zduńska Wola, Świnoujście, Żyrardów, Bolesławiec, Nowa Sól, Knurów, Oświęcim and Sopot.

Table 4.9 compiles population data for Poland's ten largest cities.

Using a few examples, describing in detail the Smart City activities that are being undertaken in selected Polish cities seems to be worthwhile. Table 4.10 lists Smart City activities and projects implemented by one of the leading cities in this field, which is Wrocław (Sokołowski, 2021).

Data presented in Table 4.10 show that Wrocław is involved in all six areas that make up the Smart City concept: Smart Mobility, Smart Economy, Smart Environment, Smart People, Smart Living and Smart Governance. Notably, in the current energy crisis, initiatives that allow for the reduction of energy consumption through the use of Smart City solutions are particularly valuable. Noteworthy in this case are solutions such as heat island metering, smart lighting systems and the SmartFlow water network monitoring system.

The city has received many national and international awards for its commitment to Smart City activities. Some of the outstanding ones are given as follows (Wrocław, 2018; Wrocław, 2021; Emitel, 2022):

• CINEV Smart Mobility in Smart City (2015) – an award received in Hong Kong for the integration of public transportation networks.

Table 4.9 Poland's largest cities

City	Number of inhabitants
Warsaw	1,860,281
Kraków	800,653
Wrocław	672,929
Łódź	670,642
Poznań	546,859
Gdańsk	486,022
Szczecin	396,168
Bydgoszcz	337,666
Lublin	334,681
Białystok	294,242

Source: List of cities (2022), Area and population (2021).

Table 4.10 Examples of Smart Cities activities implemented in Wrocław

Key area	Project	Description of initiatives
Smart Mobility	Smart Trip	The project, which has been running since 2016 and is based on an analysis of transportation resources, including roads, parking lots, rental cars, bicycles and streetcars, is expected to lead to the optimization of the use of these resources. In its advanced version, based on the traveler's preferences, the tool is expected to help him not only choose the right mode of transportation and a convenient route but also minimize the cost of the trip and facilitate payment. The Mobill app is an example of the outcome of the project.
	The Mobill App	It allows you to optimally plan a trip using public transportation and pay for it. The app locates the user with GPS coordinates and then shows the most convenient trip by public transport. The Mobill system is based on the operation of transmitters using Bluetooth technology – so-called beacons, which are deployed in streetcars and buses. An additional advantage of Mobill is access to an information guide, where users can find information about events in the city.
	Vozilla – Municipal EV Rental	At the end of 2017, 190 cars were available for users to rent, requiring only a smartphone and registration on a dedicated mobile app. Additional perks of this solution for users include free parking with Vozilla vehicles throughout the city, dedicated parking spaces in the city center and the possibility to use bus lanes.
	ITS – Intelligent Transport System in Wrocław	The system records data from road control and measurement devices (e.g., cameras and bus stop signs) and the city's public transportation, then processes the data and makes it available to traffic users. Information points provide passengers with real-time information on bus and streetcar arrivals. Based on data from the system, traffic signals can be adjusted to optimize traffic flow.

(Continued)

Table 4.10 (Continued)

Key area	Project	Description of initiatives
	Smart Parking	The goal of the project is to select the optimal solution to identify available parking spaces, suggest convenient parking options and make these data available to drivers.
Smart Economy	Startups	Wrocław is second only to Warsaw in terms of startup activity, and the city is placing increasing emphasis on support for startups. The startup database will ultimately include data on all startups that have been created or are active in Wrocław. The project will also introduce its creators and organize industry events. Young entrepreneurs will be able to count on mentoring and in the near future on special development projects.
	Open Data	Open data is information or datasets that everyone can freely use. They can be distributed and used without restriction, including for commercial purposes. Making them structured and publicly available is intended to inspire entrepreneurship, increase involvement in city affairs and demonstrate the transparency of the city's operations. The open data has been divided into categories, including transportation, sports and recreation, city hall, education or spatial data.
	Citylab	It is an initiative that offers entrepreneurs, startups and scientists the opportunity to test in the living urban fabric their original solutions and tools that positively impact the city's functioning. The priority is to acquire knowledge and explore innovative technologies that affect the environment and the health of Wrocław residents.
Smart Environment	SmartFlow	The tool, developed by MPWiK (Municipal Water and Wastewater Management) and Microsoft, allows monitoring the condition of the water supply network. With data-logging sensors distributed around the city, it is possible to pinpoint the exact location of network failures. SmartFlow also collects historical data, so problems are resolved faster and with less waste. In 2016, for example, half a billion liters of water were saved with SmartFlow.

(*Continued*)

Table 4.10 (Continued)

Key area	Project	Description of initiatives
	Wrocław EV charging system	The project, which is designed to promote environmentally friendly electric car travel, involves the construction of electric vehicle charging infrastructure. The charging network completed so far in Wrocław includes ten terminals for electric vehicles.
	'Heat Island' measurement	The project is being implemented since 2018. With a combination of sensors reporting temperature and humidity, it will be possible to determine the areas and times when Wrocław shows excessive temperatures. Leveling heat islands and improving living comfort in these areas will also be possible by planting greenery or providing water curtains.
Smart People	Wrocław rozmawia (Wrocław talks)	A platform that enables both broad public consultations and facilitates local meetings with residents or groups around specific issues. Topics may include public space management or public transportation. Also, using the website one can obtain free legal assistance or submit a petition.
	Wrocław Civic Budget	Participatory budget is influenced by all city residents. The project aims to involve Wrocław residents in the discussion of important issues or the selection of projects that change the city. Thus, residents co-determine the directions of the city's development.
	Centrum seniora (Senior Center)	A platform for the exchange of information between seniors, as well as between institutions and companies that work for the elderly. The Center runs various activation programs and projects, including Senior Days, Senior Club Federation, Senior Friendly Places and Senior Club Academy.
Smart Living	Municipal Internet	A free wireless Internet access in the city's public spaces, i.e., tourist spots associated with cultural and social life. With more than 550 access points, the Municipal Internet in Wrocław is one of the largest networks in Poland.

(*Continued*)

Table 4.10 (Continued)

Key area	Project	Description of initiatives
	LoRaWan Wireless Communication	The project is being carried out jointly with Thaumatec and Wrocław University of Science and Technology. The goal of the project is to develop a low-power radio network using LoRaWan technology, which is useful for Internet of Things (IoT) solutions. The advantages of the network are low power usage, greater security and coverage, and lower expansion costs.
	Smart Lighting	The smart lighting of public spaces uses the Philips CityTouch remote control system and LED technology, which was applied to the Nowe Żerniki housing development (WUWA2, Wohnung und Werkraum). With this solution, less energy is consumed as the intensity of lighting is adjusted to traffic and time of day.
	Edu-man	The project focuses on developing the city's telecommunications network and building a fiber-optic network connecting facilities scattered throughout the city, the operation of which is impossible without a data transmission system. Educational institutions were included in the MAN Wrocław network, and the EDU-MAN educational sub-network was created, which connects more than 250 educational institutions into one coherent data transmission system. Connection to the project is linked to lower telephone costs. In addition, the Information Services Center has purchased new computers and printers for the establishments. There are also plans to include kindergartens in the project.
	Housing Cooperative Wrocław	These are initiatives undertaken by residents who step into the role of developers and create a place to live that is 'tailored' to their needs. Individuals decide to jointly acquire a property in order to build a residential building on it. Future residents independently decide on the development of the plot, the division of space and financing. Three cooperatives have so far completed their buildings on the WUWA2 model estate in Nowe Żerniki.

(Continued)

Table 4.10 (Continued)

Key area	Project	Description of initiatives
	Open Payment	A system that uses contactless payment technology in public transportation. New ticket punchers have appeared in Wrocław streetcars and buses, allowing travelers to pay for their trip using a payment card, Urbancard EP card or smartphone. The traveler does not get a paper ticket after paying for the trip, and the cashier converts the payment card number into a token, which will be in the central system.
	Urbancard Premium	A city card dedicated for residents of Wrocław (adults and children), which can be used to pay for tickets on public transportation. Its holders can also count on a number of additional benefits in the form of attractive discounts at Partner Points participating in the program (cheaper tickets to the theater or the zoo, discounts at shopping malls, the ability to pay for parking or rent a book at Mediateka). The purpose of the program is to encourage residents to settle their income tax in Wrocław.
Smart Governance	Mobilny Asystent Mieszkańca (Mobile Resident Assistant)	An application for people involved in the life of the city, which has been in operation since September 2016. Residents can contact the relevant services in the event of malfunctioning of the city's infrastructure and can report an emergency or the need for intervention.
	Wrocław Spatial Information System	The system is designed to acquire, process and present spatial information data and accompanying descriptive information about the city's objects. Due to the fact that the system is updated on an ongoing basis, it not only forms the basis for city administration activities but also provides interested parties with knowledge about urban space.
	ePUAP – Electronic Platform of Public Administration Services	A tool that connects residents of Wrocław, entrepreneurs and institutions to an online access to public administration services. Through the platform it is possible to remotely submit complaints, applications, inquiries to the office, change the address of correspondence or residence.

(*Continued*)

Table 4.10 (Continued)

Key area	Project	Description of initiatives
	Wrocław Information and Payment Platform	The tool, using a user profile in ePUAP, facilitates checking and settling online tax liabilities and selected civil law fees, such as fees for perpetual usufruct of land or lease collected by the municipality of Wrocław.
	E-Application	The solution eliminates the need for several visits to the office to submit the application and pick up the requested license plate, and, in the case of residents of surrounding counties, the long waiting time for the Wrocław City Hall to process the application for assigning a distinguishing code for the vehicle.
	Smart Resident Information Management	The project aims to collect data based on a satisfaction survey available to residents of Wrocław and tourists visiting the city. The research allows determining the level of satisfaction regarding life in Wrocław and the use of city infrastructure.

Source: Wrocław (2018), Smartcity Wrocław (2022), Smartcity Wrocław Projects (2022), Bednarska-Olejniczak et al. (2019).

- City of the Year over 500,000 residents (2016) – a statuette presented at the Smart City Forum for the vision of building Wrocław as a Smart City based on pillars such as strategy, residents and communication with them, the attractiveness of life and development and creativity, manifested, among other things, in the opening of data and promotion of the startup community.
- Public incentives in transportation (2016) – an award presented at the 2016 Euro-China Smart Mobility Conference in Shenzhen for modern transportation solutions that encourage city residents to travel by public transportation.
- City of the Year over 500,000 residents (2018) – an award received at the Smart City Forum for innovative solutions in the field of electromobility: Municipal EV Rental 'Vozilla' and cashless city smart payments.
- Green & Smart City Awards (2018) – an award presented at the global mart City Expo summit in China in the Top Level Design category for a mobile app comprising 15 projects that use smart solutions in urban spaces and serve to improve the lives of residents and two projects related to ecology.

- First place in the 'Cashless Cities' ranking (2021) – the award went to Wrocław for the city's computerization and smart solutions, in which cashless transactions dominate.
- Emitel awarded in the Smart City Awards in the Smart City Solutions (2022) category – the awarded solution is the construction of an innovative system for remote reading of parameters from water meters for the Municipal Water and Wastewater Management (MPWiK). This is the first municipal Internet of Things (IoT) project of this scale in Poland, which will eventually include up to more than 70,000 meters. The delivered solutions will also provide enhanced connectivity to devices located in hard-to-reach locations, such as deep wells or basements.

In particular, it is worth taking a broader look at issues concerning startups, a very important factor in the economic development of the modern smart city. In the case of Wrocław, the city has for years pursued an active investment policy aimed at attracting Polish and foreign companies targeting technology projects. When it comes to supporting startups, Wrocław plays four major roles (Brdulak, 2017; Brzuśnian, 2017; Wrocław, 2018):

- As the creator of modern research infrastructure (Wrocław Technology Park or Polish Center for Technology Development – formerly Wrocław Research Center EIT Plus);
- As a customer/recipient of solutions created by local startups (e.g., Explain Everything educational app in Wrocław schools);
- As a partner in a startup accelerator program (MIT Enterprise Forum Poland);
- As an initiator of projects aimed at integrating the startup community (the activities of the Startup Wrocław team).

The city has engaged in numerous initiatives to support the development of startups. The Wrocław Agglomeration Development Agency has been established to provide comprehensive assistance to the development of local startups. The activities of the said agency focus on the following aspects:

- Networking – integration of the startup ecosystem in Wrocław;
- Events – organization of technology and startup events;
- Startup base – information service https://www.wroclaw.pl/startupy/ and the unique database of Wrocław startups located there;
- Support and promotion – support in dealing with large companies investors, media and business environment institutions.

Another example of a Polish city heavily involved in the implementation of Smart City solutions is Lublin. Table 4.11 lists examples of Smart City projects implemented in Lublin. In this case, it can be seen that the city's activities are concentrated in five areas of Smart City: Smart Mobility, Smart Environment, Smart People, Smart Living and Smart Governance. The city of Lublin is the most involved in Smart Mobility issues – in this case, in recent years, a number of different initiatives and projects have been implemented on car sharing, urban bicycles and better use of urban space, improving public transportation and improving mobility in the city. Also, the city is strongly involved in Smart Living initiatives in this case paying attention to the needs of digitally excluded people such as seniors, trying to engage residents in the form of participatory budgets, has introduced a city card as well as creating various platforms for sharing services among residents.

Table 4.11 Examples of Smart Cities activities implemented in Lublin

Key area	Project	Description of initiatives
Smart Mobility	Lublin City Bike	The aim of the project was to improve the tourist attractiveness of the region by creating an unmanned network of bicycle stations, increasing accessibility to tourist and recreational infrastructure, and increasing the positive environmental aspect. The project is based on a system of urban bicycle rentals. The stations are located throughout the city of Lublin. Using the system is simple, as bicycle rental is done via the NextBike mobile app or by entering the bicycle number as a QR code into the rental terminal.
	Mobile Sharing Services	Creating a friendly urban space, including reducing congestion and improving air quality, requires changes in thinking and perceptions of everyday mobility. This is to be served by the concept of so-called shared mobility with regard to both sharing cars and micromobility (scooters, motor scooters and bicycles).
	Low-carbon Public Transport Network	This project's material scope includes purchasing ten 18 m trolleybuses with additional propulsion, 20 electric buses and 14 buses meeting EURO6 emission standards. In addition, the city also plans to expand the dynamic bus stop information system on the roads covered by the project and integrate them into the currently used system, as well as to build trolleybus tractions.

(Continued)

Table 4.11 (Continued)

Key area	Project	Description of initiatives
	CITYHON	The project's main goal was to organize an urban hackathon with a theme related to urban mobility in its broadest sense. The ambition of the project is also to create a reference event in Europe, which will be tasked with crushing top talent and creating innovative mobility solutions.
	AI-TraWELL	The project's main objective is to develop a prototype tool based on artificial intelligence (AI), which, in the form of an interactive chatbot, will recommend personalized travel routes, taking into account the diverse needs and preferences of users. The AI-TraWELL prototype app under development is aimed at combining users' needs and preferences with real-world information enriched with predictions on available transportation modes. The AI-TraWELL app is distinguished by the fact that it aggregates all available means of transportation in the city, taking into account not only statistical data but also subjective data defined by the application's user, while enabling precise delivery of individualized information.
	Multistage Design Thinking Project Support	The goal of the project was to create more sustainable solutions for urban mobility by training public officials responsible for this subject area using the design thinking method. Design thinking as a set of cognitive and practical processes is widely used and popular among startups, creative industries and private companies because of its effectiveness in solving complex problems whose solutions go beyond the application of technical knowledge.
Smart Environment	Water and Wastewater Network Management: a mathematical network model	An important part of the project is the expansion of the monitoring system for technological measurements (pressures, flows, sanitary sewer filling status and sewage quality). The heart of the Central Control System will be a SCADA system, allowing visualization and control of the entire water production process and programming of equipment operation.

(*Continued*)

Table 4.11 (Continued)

Key area	Project	Description of initiatives
	Remote water meter reading with data acquisition for Municipal Water and Wastewater Management (MPWiK Sp. z o.o.).	The project includes the installation of water meter overlays with telemetry modules equipped with SIM cards with assigned IP addresses. The telemetry modules perform cyclic data recording (water meter status) and periodically send a set of data to a web application installed on the MPWiK server. The application allows access to these data and, additionally, its analysis (e.g., alert on water outages).
	Implementing the operation optimizing program for the heating network	To increase the optimization of district heating network operation, a system approach was designed. The system consisted of three functional modules: a Offline Module – It allows to build and calibrate, based on SCADA systems data, a mathematical model reproducing as accurately as possible the physical parameters of district heating system components. b Online Module – This is a calculation module for real-time operation, simulation of the state in fixed defined time steps (e.g., hourly) on the basis of data downloaded on an ongoing basis from SCADA systems, constant verification of the obtained calculation results with actual data downloaded from telemetry systems. c Temperature Optimizer – The most innovative module that optimizes the efficiency of operation by dynamically calculating in a continuous mode (with a complex time step of, for example, one hour) the optimal supply parameters so that all consumers receive the medium at the correct temperature.
	ekoAPP	ekoAPP is a mobile application developed for residents of Lublin. The software is designed to help with waste management. The ekoAPP user will be able to check the collection date for each waste fraction, along with an automatic reminder option. EkoAPP will allow adding several locations, so users will be able to receive information for several addresses in one place. The app also supports the process of selective waste collection by answering questions related to assigning a given waste to the correct fraction.

(Continued)

Table 4.11 (Continued)

Key area	Project	Description of initiatives
Smart People	Let's invent Lublin together	The new concept of the strategic document assumes that the key idea and axis of the whole process is precisely the social participation and involvement of the city's citizens in the largest and most diverse group possible. The process will consist in joint work on the formulation of the Lublin 2030 Strategy document with the help of various participatory techniques and methods selected appropriately to the selected thematic area so that they are as effective and widely available as possible, avoiding limitations such as competence.
	Otwarte Dane miasta Lublin (City of Lublin Open Data)	The City of Lublin Open Data portal is a constantly expanding repository of knowledge based on data created and processed by the city government and its subordinate units. Recipients using public data, now and later, are all those interested in the use of public data, i.e., both individuals and businesses, public administration units, non-governmental organizations, entities of science, culture and others.
	Lublin3D – Spatial Information System	Lublin 3D is an interactive map of Lublin extending the functionality of Geoportal with additional information presented thanks to the '3D' feature. It allows the easier conception of space, showing the results of complex analyses, conducting discussions on topics that require an interactive presentation. What is more, it allows downloading data in software formats dedicated to 3D files – thus facilitating the work of designers and pupils or students. It is a database of information (not counting data from the city's Geoportal) on building heights, obscuration, shadow analysis and cross-sections through the terrain. It is also a pre-design tool visualizing conceptual development or designed objects.

(*Continued*)

Table 4.11 (Continued)

Key area	Project	Description of initiatives
Smart Living	SOS for Seniors	Coordinated pilot service for people who need constant monitoring and support and those with health risks who want to increase their independence. It provides a sense of security for both seniors and their loved ones who are not around on a daily basis. It not only allows monitoring basic life activities and alarm events, a fall or leaving a safe location but also gives the senior the ability to communicate with the Telecare Center. Each participant in the 'SOS for Seniors' program receives a wristband with an SOS (emergency) button and a SIM card to be able to make a voice call to the Telecare Center.
	Citizens' Panel – Lublin 2018: What to do to breathe clean air?	A citizen panel is a technique for making important decisions for the city with the participation of male and female residents. It is a form of deliberative democracy. Several factors contribute to its specificity.
	Green Participatory Budgeting	Lublin is the first city in Poland to introduce a Green Budget. Every resident of the city can now influence the creation of new green areas or the revitalization of existing ones. The program is designed to improve the quality of life of residents and the functionality of green areas located in our city. The implementation of the Green Budget is supervised by the Office of the Municipal Greenery Architect, supported by other departments of the Lublin City Hall and the Roads and Bridges Administration. Submitted ideas for the development of Lublin's space in the initial stage undergo a formal evaluation, which includes verification of, for example, ownership of the land on which the projects are located and the possibility of implementation within one budget year.
	Lublin City Card – LUBIKA	The Lublin City Card is one of the components of the new Lublin Ticket Agglomeration Card (LUBIKA) system, which is developing modern solutions and charging methods. The city card itself is a loyalty program for Lublin residents. Ownership of the Lublin City Card entitles residents to discounts, rebates and concessions on city services and products offered by partners in such areas as public transportation, Paid Parking Zone, culture, sports, recreation, education, health, catering, trade and services.

Table 4.11 (Continued)

Key area	Project	Description of initiatives
	Naprawmy To Platform	A service that allows its users to report problems observed in their immediate surroundings in public space. The basic function of the service is to mark the location of the problem on a map, add a description of it and the possibility to inform about the report the appropriate institutions responsible for solving specific categories of issues. The issues themselves are divided into categories such as infrastructure, security, nature, buildings and others. The tool allows users to quickly and efficiently identify problems arising in different areas of the city's functioning.
	Lublin Virtual Library	The project design included the construction of network, server and database infrastructure at the partners and participants of the investment, and a digitization studio. As a result of the project, a fiber optic network infrastructure was created to connect the partners' ICT resources with the network of the Lublin municipality. The key element of the Lublin Virtual Library (LBW) is a central portal with separate instances for the partners. The LBW portal's search engine helps develop, narrow and specify search results and offers smart filters, hints and suggestions.
	Lublin Free WiFi	The project was submitted through the 2014 Citizens' Budget. After the project was positively evaluated, it was put to the vote of residents. Free WiFi in Lublin received the highest number of votes, which meant that work began on the design of the city network. Throughout the implementation process, there was cooperation between the person submitting the project, residents, the City Hall and the Department of Information Technology and Telecommunications. The process of determining the location and distribution of devices was consulted with the residents themselves, allowing the deploying of access points in the most active public spaces.

(*Continued*)

Table 4.11 (Continued)

Key area	Project	Description of initiatives
	Smart city benches	As part of this measure, three smart benches were installed in the city of Lublin in the Bystrzyca River areas. The solar benches are equipped with two inductive chargers and two USB outputs to charge a mobile device such as a phone, tablet or laptop. The benches also provide free Wi-Fi. The smart benches are eco-friendly – they do not need external power. The built-in batteries are charged by solar modules. The bench can be used both during the day and at night. Also, trash cans equipped with sensors that inform city services when they are overflowing and need to be emptied were installed next to the benches.
Smart Governance	Lublin Municipality E-services – construction and expansion	A project to expand the Spatial Information System, introduce new e-services for residents and entrepreneurs, and organize the range of data available at the Lublin City Hall and make it available to a wide audience: a Participatory portal – This portal allows the possible active participation of Lublin residents. The system will allow participation and access to all procedures and tools concerning: Civic Budget, Green Budget, local initiatives, public consultations and submission of applications/feedback. The panel will allow for the submission of projects for the Civic Budget, applications under the local initiative, as well as applications and comments on local plans and the study of spatial planning conditions and directions. b Open Data Module – This is a portal from which residents and stakeholders will be able to download data from databases of the Lublin City Hall in a form easy for further processing, analysis and use in information systems. The whole will be compatible with the national data platform https://danepubliczne.gov.pl/

Source: Lublin, urban innovations (2022), Lublin Digital Transformation (2022), Horosiewicz (2021), Bojanowska and Lipski (2019), and Guzal-Dec et al. (2019).

As in the case of Wrocław, Lublin has received awards for its involvement in the Smart City area. Some of them include the following (Laureates, 2019; Lublin 2020):

- Smart City Award (2020) – awarded for increasing the accessibility and transparency of digital tools, which demonstrates the implementation of a real strategy for city development in accordance with the Human Smart Cities paradigm in the category of 100,000–500,000 residents.
- Smart City Award (2020) – awarded for implementing a comprehensive spatial model of the city. The award was presented at the 11th Smart City Forum in the category of 100,000–500,000 residents.

The case studies of selected cities in Central and Eastern Europe, including especially Poland, discussed in this chapter allow us to conclude that the region's cities are strongly engaged in the implementation of the Smart Cities concept. Much remains to be done in this regard, as the concept is not evenly implemented in all cities (as will be shown in the following chapters). However, in all the countries analyzed, one can find numerous cases of good practice in the implementation of the Smart City concept, on which other cities can model themselves.

Bibliography

Action plan for the City of Prague. (2020). https://projects2014-2020.interregeurope.eu/fileadmin/user_upload/tx_tevprojects/library/file_1587390320. pdf, [access data: 17.11.2022].

Area and population in the territorial profile in 2021. https://stat.gov.pl/en/ topics/population/population/area-and-population-in-the-territorial-profile-in-2021,4,15.html, [access data: 17.11.2022].

Baculáková, K. (2019). Solutions for a smart urban environment air quality, transport and energy systems: Knowledge from Helsinki and lessons to be learned for Bratislava. *Geopolitics of Energy, 41*(2), 8–15.

Baculáková, K. (2020). Selected aspects of smart city concepts: Position of Bratislava. *Theoretical and Empirical Researches in Urban Management, 15*(3), 68–80.

Bednarska-Olejniczak, D., Olejniczak, J., Svobodová, L. (2019). Towards a smart and sustainable city with the involvement of public participation-The case of Wroclaw. *Sustainability, 11*(2), 332.

Bojanowska, A., Lipski, J. (2019). The use of data by smart systems for price forecasting in the context of building customer relationships on the Lublin real estate market. *IOP Conference Series: Materials Science and Engineering, 710*(1), 012001.

BRATISLAVA rozumné mesto 2030. (2018). https://bratislava.blob.core.windows.net/media/Default/Dokumenty/smartcity%20rozumna%20bratislava2030.pdf, [access data: 17.11.2022].

Brdulak, A. (2017). Idea Smart City w strategii zarządzania urzędem miasta we Wrocławiu. *Studia Miejskie, 27*, 143–154. htps://doi.org/10.25167/sm2017.027.11

Brzuśnian, A. (2017). Wrocław jako przykład miasta inteligentnego. *Research Papers of Wrocław University of Economics, 470*, 29–39.

Budapest Smart City Centre of Excellence. (2015). https://cordis.europa.eu/project/id/664605, [access data: 17.11.2022].

Czech Smart City Cluster. (2022). https://czechsmartcitycluster.com/en/about-the-cluster/, [access data: 17.11.2022].

Dudzevičiūtė, G. Šimelyt, A., Liučvaitienė, A. (2017). The application of smart cities concept for citzen of Lithuania and Sweden: Comparative analysis. *Independent Journal of Management and Production, 8*(4), 1433–1450.

Emitel laureatem Konkursu Smart City Awards w kategorii Smart City Solution. (2022). https://emitel.pl/aktualnosci/emitel-laureatem-konkursu-smart-city-awards-w-kategorii-smart-city-solution/, [access data: 17.11.2022].

Establishment of the Budapest Smart City Centre of Excellence at the Budapest University of Technology and Economics. (2015). https://smartpolis.eit.bme.hu/, [access data: 17.11.2022].

Golej, J., Panik, M., Adamuscin, A. (2016a). Smart Infrastructure in Bratislava, ICST Institute for Computer Sciences. *Social Informatics and Telecommunications Engineering*, 142–149, https://eudl.eu/pdf/10.1007/978-3-319-47075-7_17, [access data: 17.11.2022].

Golej, J., Panik, M., Adamuscin, A. (2016b). Smart infrastructure in Bratislava. *Lecture Notes of the Institute for Computer Sciences, Social-Informatics and Telecommunications Engineering, LNICST, 170*, 142–149.

Guzal-Dec, D., Zwolińska-Ligaj, M., Zbucki, L. (2019). The potential of smart development of urban-rural communes in peripheral region (a case study of the Lublin Region, Poland). *Miscellanea Geographica, 23*(2), 85–91.

Horosiewicz, Sz. (2021). *Raport Doświadczenia miasta Lublin z zakresu projektów o tematyce mobilności miejskiej*, https://comobility.edu.pl/wp-content/uploads/2021/12/Artykul_Lublin_z.pdf, [access data: 17.11.2022].

Laureaci Konkursu Smart City za rok 2019. (2019). https://smartcityforum.pl/laureaci-konkursu-smart-city-2019/, [access data: 17.11.2022].

List of localities in alphabetical order. (2022). https://www.ksh.hu/apps/hntr.egyeb?p_lang=EN&p_sablon=TELEPULESEK_ABC, [access data: 17.11.2022].

Lista miast w Polsce. (2022). https://www.polskawliczbach.pl/Miasta, [access data: 17.11.2022].

Lithuania, An open platform for public sector transformation. (2022). https://lithuania.lt/governance-in-lithuania/smart-city-solutions/, [access data: 17.11.2022].

Lublin 2030. Europejska metropolia? (2018). Urząd Miasta Lublin, Lublin, https://depot.ceon.pl/bitstream/handle/123456789/16243/Lublin%202030%20-%20wersja%20cyfrowa.pdf?sequence=1&isAllowed=y, [access data: 17.11.2022].

Lublin z nagrodą Smart City Award. (2020). https://metropolie.pl/artykul/lublin-z-nagroda-smart-city-award, [access data: 17.11.2022].

Lublin, Innowacje miejskie (2022). https://smartcity.lublin.eu/smart-city-lublin/innowacje-miejskie/, [access data: 17.11.2022].

Pacovský, L., Jolič, J. (2021). Smart cities and the necessity of opening of the data in the Czech republic as an example of CEE country. *EMAN 2021 Conference Proceedings, The 5th Conference on Economics and Management*, 129–137. htps://doi.org/10.31410/EMAN.2021.129, [access data: 17.11.2022].

Peterka, J. (2019). *Smart Prague and the City's energy savings*, https://eu.eventscloud.com/file_uploads/41e1c677e91483679e5c9f14dd79a5ab_SP_Energetika_Presentation_finalbyJIRKAPETERKA.pdf, [access data: 17.11.2022].

Population of Cities in Lithuania 2022. (2022). https://worldpopulationreview.com/countries/cities/lithuania, [access data: 17.11.2022].

Population of Cities in Slovakia 2022. (2022). https://worldpopulationreview.com/countries/cities/slovakia, [access data: 17.11.2022].

Population of Municipialities. (2022). Chech Statistical Office, Pardubice, https://www.czso.cz/csu/czso/population-of-municipalities-1-january-2022, [access data: 17.11.2022].

Powierzchnia i ludność w przekroju terytorialnym. (2021). Główny Urząd Statystyczny, https://stat.gov.pl/obszary-tematyczne/ludnosc/ludnosc/powierzchnia-i-ludnosc-w-przekroju-terytorialnym-w-2021-roku, 7,18.html, [access data: 17.11.2022].

Projekty smartcity Wrocław. (2022). https://www.wroclaw.pl/smartcity/projekty, [access data: 17.11.2022].

Rohlena, M., Frkova, J. (2014). *Smart cities – pilot project Smart Prague*, Construction Maeconomics Conference 2014, http://www.conference-cm.com/podklady/history5/Prispevky/paper_Rohlena.pdf, [access data: 17.11.2022].

Smart Budapest. (2017). Municipiality of Budapest, https://budapest.hu/Documents/V%C3%A1ros%C3%A9p%C3%ADt%C3%A9si%20F%C5%91oszt%C3%A1ly/Smart_Budapest_summary_ENG.pdf, [access data: 17.11.2022].

Smart City Vilnius. (2022). http://www.smartcityvilnius.com/en/home.html, [access data: 17.11.2022].

Smart Prague – Vision 2030. (2022). https://www.smartprague.eu/en, [access data: 17.11.2022].

Smartcity Wrocław. (2022). https://www.wroclaw.pl/smartcity/, [access data: 17.11.2022].

Sokołowski, J. (2021). Rola kapitału ludzkiego w koncepcji Smart City na przykładzie Wrocławia. In: Zakrzewska-Półtorak, A. (eds) *Debiuty Studenckie red.* Wydawnictwo Uniwersytetu Ekonomicznego we Wrocławiu, 22–41.

Štěpánek, P. (2020). Smart city through design: Preparation of a new wayfinding system in Prague. *Lecture Notes in Computer Science (including subseries Lecture Notes in Artificial Intelligence and Lecture Notes in Bioinformatics), 12423 LNCS*, 514–526.

Svítek, M., Dostál, R., Kozhevnikov, S., Janča, T. (2020). Smart City 5. 0 Testbed in Prague, *2020 Smart City Symposium Prague (SCSP)*, 1–6. htps://doi.org/10.1109/SCSP49987.2020.9133997

Szegvári, P. (2017). *Smart city activities of Budapest, regarding the EU Urban Agenda and JASPERS-related activities*, http://www.jaspersnetwork.org/download/attachments/23364212/5.Budapest%20-%20Smart%20city%20activities.pdf?version=1&modificationDate=1495466079000&api=v2, [access data: 17.11.2022].

Turoń, K., Sierpiński, G., Tóth, J. (2022). Support for pro-ecological solutions in smart cities with the use of travel databases – A case study based on a bike-sharing system in Budapest. *Advances in Intelligent Systems and Computing, 1091 AISC*, 225–237.

Vácha, T., Kandusová, V. (2018). Making innovation in elderly care possible using participatory design: The smart home-care project in Prague. *2018 Smart Cities Symposium Prague, SCSP 2018*, 1–6.

Wrocław w kierunku Smart City. (2018). Agencja Rozwoju Aglomeracji Wrocławskiej SA, Wrocław.

Wrocław z trzema nagrodami. (2021). https://www.wroclaw.pl/przedsiebiorczy-wroclaw/local-trends-2022-wroclaw-z-nagrodami-smart-city-cashless-biznes, [access data: 17.11.2022].

Zapolskytė, S., Trépanier, M., Burinskienė, M., Survilė, O. (2022). Smart urban mobility system evaluation model adaptation to Vilnius, Montreal and Weimar Cities. *Sustainability, 14*(2), 715.

5 Research intentions and assumptions

5.1 Research intentions, problems and sample selection

The remainder of this monograph focuses on the practical problems of operation of cities aspiring to the title of Smart City located in Central and Eastern Europe. The considerations devoted to these cities are divided into several key areas that, on the one hand, are related to the issue of sustainability (Mora et al., 2021; Su and Fan, 2022), and, on the other hand, refer to the next stages in the development of smart cities and the systematic expansion of cooperation between local authorities and urban stakeholders (business, science, community, environmental organizations) (Komninos et al., 2019). With these aspects in mind, the next four chapters will respectively address the issues:

- economic and financial defining framework for the development and civilization of urban communities;
- infrastructural and technical defining basic determinants of quality of life in cities;
- social and demographic determinants of communal and non-communal conditions of urban existence;
- environmental issues related to respect for natural resources for their rational use by present and future generations.

The chapters have a uniform internal structure. They begin with a subsection providing an overview of data on the determinants of urban development (as defined by the chapter's subject matter) in Central and Eastern Europe. The next two subsections contain the results of research conducted in Polish cities in terms of a) quantitative (analysis of statistical data) and b) qualitative (analysis of survey results).

The first of the above-mentioned research layers uses a general, overview perspective based on literature studies and case studies. In the second aspect of the research, an analysis is carried out of statistical data on the 16 largest Polish cities (provincial cities), most of which aspire to be

DOI: 10.4324/9781003358190-5

smart, and all of which are trying to implement smart solutions in infrastructure and local life. The third research layer presents the results of surveys conducted on a representative sample of 287 Polish cities. They make it possible to learn about the subjective assessment of the analyzed aspects made by representatives of city authorities responsible for implementing urban development strategies.

Table 5.1 shows the characteristics of the Polish cities analyzed in the second research layer. Figure 5.1 illustrates their geographical distribution.

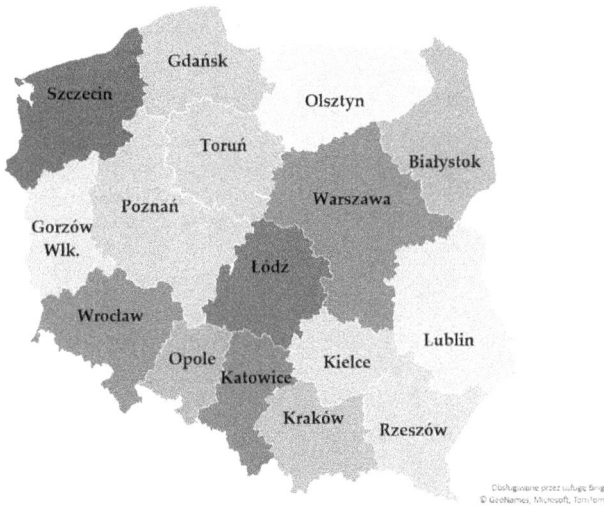

Figure 5.1 Distribution of the surveyed provincial cities on the map of Poland.

Source: Own compilation using Microsoft Excel map.

Table 5.1 Characteristics of the 16 surveyed provincial cities (as of 31/12/2022)

City	Population	Surface	Population density	Industry
Białystok	296,000	102 km²	2,902 persons/ km²	Electro-mechanical (electronics, machinery and metal), wood, clothing, food and printing industries
Gorzów Wlk.	120,087	86 km²	1,400 persons/ km²	Chemical, energy, electronics, light, machinery, pharmaceutical, automotive, metal and food industries

(*Continued*)

Table 5.1 (Continued)

City	Population	Surface	Population density	Industry
Gdańsk	471,000	263 km²	1,787 persons/ km²	Shipbuilding, petrochemicals, energy, apparel and metals
Katowice	292,000	165 km²	1,756 persons/ km²	Mining, business services and automotive
Kielce	192,500	110 km²	1,686 persons/ km²	Construction, building materials, electrical machinery, as well as food and processing industries
Kraków	782,000	327 km²	2,450 persons/ km²	Tourism, business services, trade and banking services
Lublin	338,000	147 km²	2,270 persons/ km²	Energy, chemical, food and tobacco
Łódź	670,642	293 km²	2,287 persons/ km²	Banking service centers, household appliance manufacturing and pharmaceutical industry
Olsztyn	170,622	83 km²	1,932 persons/ km²	Tire, wood and furniture, meat, dairy, milling, clothing, building materials and printing industries
Opole	127,839	149 km²	858 persons/ km²	Building materials, food, machinery and equipment and IT industries
Poznań	532,000	262 km²	2,031 persons/ km²	Electromechanical, chemical, commercial and transportation
Rzeszów	198,609	129 km²	1,539 persons/ km²	Aerospace, household appliances, machinery and pharmaceutical industry
Szczecin	396,472	301 km²	1,319 persons/ km²	Maritime, shipbuilding and electrical installation industry
Toruń	197,812	116 km²	1,511 persons/ km²	Pharmaceutical, cosmetic, food, electro-technical and metal industries
Warsaw	517,000	517 km²	3,466 persons/ km²	Electrical engineering, transportation equipment, chemical, food and printing
Wrocław	643,000	293 km²		Machinery, transportation equipment, food, electro-technical, metal, clothing and chemical industries

Source: Own work.

In the course of the research and in the process of inference, particular attention was paid to two cities: Warsaw and Wrocław (Orłowski, 2021; Smętkowski et al., 2021). These are the entities that are most often mentioned as Polish smart cities in international rankings of smart cities. They were also included – as the only ones – in the prestigious *IESE Cities in Motion Index*. In the group analyzed, these cities were compared to others so that the potential distance separating them from other provincial cities could be assessed.

As already mentioned, the third research layer analyzed the results of surveys conducted on a representative sample of Polish cities comprising 287 entities from 930 existing cities in Poland. In the process of determining the sample size, a 95% confidence level was assumed, which means that the results obtained reflect 95% of the actual state. It was also assumed that the value of the maximum error may be 5%, so the indications of respondents may differ by ±5% from the values calculated for the entire population.

The characteristics of the cities that were eventually included in the survey sample are presented in Table 5.2. It takes into account two key metric questions relating, respectively, to the size of the city expressed in terms of population and the economic situation defined in terms of per capita budget income. Thus, the largest number of respondents represented cities with a population of 10,001 to 25,000 and a budget income of PLN 4,001 to PLN 5,000. The share of the remaining units in the sample was fairly evenly distributed, reflecting the actual distribution of Polish cities.

Table 5.2 Characteristics of the survey sample of cities surveyed

Size of the city expressed in terms of population		
Below 5,000 residents	41	14.29%
5,001–10,000	40	13.94%
10,001–25,0000	83	28.92%
25,001–50,000	55	19.16%
From 50,001 to 100,000 residents	34	11.85%
Over 100,000 residents	34	11.85%
Total	287	100.00%
The economic situation of the city as expressed by the amount of budget income per capita		
PLN 1,000–2,000	49	17.07%
PLN 2,001–3,000	49	17.07%
PLN 3,001–4,000	62	21.60%
PLN 4,001–5,000	82	28.57%
Over PLN 5,001	45	15.68%
Total	287	100.00%

Source: Own elaboration based on the results of questionnaire surveys.

The main research objective – in all the perspectives defined above – is to assess the quality of life of city dwellers in Central and Eastern Europe with a particular focus on Polish urban communities. This assessment is carried out in the context of the development of the Smart City concept and the sustainability of smart cities (Bhattacharya et al., 2020; van den Buuse et al., 2021). This is because the literature study shows that such a defined and extensive research topic has not been undertaken before, and the originality of the presented considerations is a direct result of the following circumstances:

- to focus research problems around issues related to the quality of urban life, which is addressed far less often in the literature and research;
- to record the assessment of quality of life in the areas of sustainable development (economic, technological-infrastructural, social, environmental);
- to take a look at the development of smart cities in emerging and developing economies located in Central and Eastern Europe;
- to conduct research on a representative sample of Polish cities, allowing to verify in practice many opinions and views related to urban life, including, above all, those related to unsustainability and reserving the Smart City concept only for large or financially prosperous cities.

In the course of considerations and research – in relation to the research gap identified above – answers are also sought to the following research problems:

- How can the Smart City concept support the improvement of the residents' quality of life?
- What characteristics and challenges characterize modern cities aspiring to be Smart (Cugurullo, 2018)?
- What conditions does a Smart City have to meet to be a sustainable city?
- What is the scale of unsustainability of the examined cities in the economic, technological, social and environmental area?
- Whether and to what extent the quality of life in the examined cities is influenced by:
  a their economic and financial situation;
  b size expressed in the number of residents.
     In the course of considerations and research – in relation to the research gap identified above – answers are also sought to the following research problems:
- How can the Smart City concept support the improvement of the residents' quality of life?

- What features and challenges characterize modern cities aspiring to be Smart?
- What conditions does a Smart City have to meet to be a sustainable city?
- What is the scale of unsustainability of the examined cities in the economic, technological, social and environmental area?
- Whether and to what extent the quality of life in the examined cities is influenced by:
  a their economic and financial situation;
  b size expressed in the number of residents;
  c regional geographic location defined by province.

5.2 Research stages and methods

The research – following the philosophy outlined in the previous section – involved three stages, and their description along with a list of the research methods used is presented in Table 5.3. These stages were ordered according to the principle of hierarchization of content from the most general to the most specific.

Table 5.3 Characteristics of research stages along with research methods

Stage	Objective	Methods
General overview of the determinants of smart city development in Central and Eastern Europe	Identification of the level of sophistication of the development of smart urban solutions in Central and Eastern Europe, taking into account the following areas: economic, technical-infrastructural, socio-demographic and environmental.	International literature review and multiple case studies of selected cities in Central and Eastern Europe.
Analysis of statistical data on the development of 16 Polish provincial cities	Evaluation of selected development parameters of Polish cities in the context of statistical data, in the following areas: 1 economic (budget revenues per capita [in PLN]; property expenditures per capita [in PLN]; budget deficit/surplus in relation to total budget revenues [in %]); 2 technical and infrastructural: share of transportation and communications spending in total budget spending [in %]; share of housing spending in total budget spending [in %];	Geographically and inter-area comparative analysis oriented to determine the level of unsustainability.

(*Continued*)

Table 5.3 (Continued)

Stage	Objective	Methods
	3 socio-demographic: percentage and change of people aged 65 and older in the total population [in %]; birth rate; net enrollment rate – elementary school [in %]; 4 environmental: expenditures on air and climate protection per capita [in PLN]; share of parks, green spaces and neighborhood green spaces in total area [in %]; share of green space in total area [in %]; share of mixed waste in total waste [in %];	
Survey on a representative sample of 287 Polish cities	The assessment of the various conditions determining residents' quality of life from the city government's perspective in the following areas: economic, technical-infrastructural, socio-demographic and environmental. The assessment was carried out on a five-point Likert scale, and the layout of the questionnaire containing each group of questions is characterized later in this chapter.	Survey results and measures of descriptive statistics (dominant, median, mean, coefficient of variation and standard deviation). Spearman's rank correlation coefficient, which is a non-parametric statistical test, was used to identify the relationship between a city's size and its economic situation. The significance of the relationships found was tested at two levels of significance: $p<0.01$ and $p<0.05$.

Source: Own work.

In the course of statistical research relating to 16 Polish provincial cities, the focus was on key indicators determined on the basis of real data, thus objectively illustrating the level of quality of life in individual thematic areas related to the sustainable development of smart cities. These data in Poland are collected and published in the form of the Local Data Bank (https://bdl.stat.gov.pl/bdl/start).

Ultimately, the choice of indicators for statistical research was based on two considerations: compatibility with the subject matter of the considerations undertaken in this monograph, and the availability of data to assess particular aspects of the quality of urban life (Wolniak and Jonek-Kowalska, 2021; Chen, 2022). It is also worth mentioning that in the search for appropriate metrics, the international standard *PN-ISO 37120 Sustainable development of communities. Indicators for city services and quality of life* (Deng et al., 2017; Pineo et al., 2018) was used.

Thus, within the framework of **economic determinants** (Monfaredzadeh and Berardi, 2015; Batabyal and Nijkamp, 2019), a city's current financial situation was assessed on the basis of budget income per capita. This is a measure used in comparative analyses of urban wealth. It is also used by government authorities in decisions regarding the allocation of subsidies and grants from the central budget. Moreover, bearing in mind the need to take into account the development perspective of the cities under study, the amount of property expenditures per capita was used in the course of economic research, illustrating the possibility of expanding the city's infrastructure, as well as the ratio of budget deficit/surplus to total income reflecting the scale of the financial gap and the risk of increasing debt.

Regarding **technical and infrastructural considerations**, the focus was on two key aspects of urban life that are very important to city residents and therefore their main stakeholders. The first aspect was transportation and connectivity, which are inextricably linked to the genesis of the Smart City concept (So et al., 2020; Dohn et al., 2022; Wang et al., 2022; Wawer at al., 2022). The second aspect was housing infrastructure, the accessibility and quality of which are a guarantee of living comfort for current and future generations (Jonek-Kowalska, 2022). For these reasons, the share of transportation and communications expenditures and the share of housing expenditures in total expenditures of the studied cities were used in the analyses of the indicated determinants. This made it possible to assess and compare cities in terms of the scale of measures taken to develop the transportation and communications network and housing infrastructure.

Socio-demographic determinants were set to address three aspects: population growth, aging of urban communities and primary education. The aforementioned aspects are a key to the reproduction, age structure and education of urban residents. The results obtained in these dimensions make it possible to determine the direction of smart city needs (Del-Real et al., 2021) and identify the scale of potential socio-demographic problems. This, in turn, allows for more effective development of smart city strategies (Kim et al., 2022; Micozzi and Yigitcanlar, 2022).

The last group of determinants included **environmental factors**, important not only for the quality of life in the city but also relevant from a global perspective and intensifying climate risks (Su et al., 2021). The level of per capita spending on environmental and climate protection was used to assess the city government's commitment to environmental protection.

This framing allows comparing cities among themselves and determining the rank of these expenditures in city budgets. In addition, the share of green areas, including those created by the city (parks, greens, neighborhood greenery) in the total area of the surveyed cities, was taken into account within the framework of the described conditions. This is because theoretical considerations on smart cities show that the development of green recreational areas is an indicator of being smart significantly improving the quality of life of residents. Smart city structures, therefore, cannot just be urban bedrooms equipped with gated residential areas. They must create a friendly and clean living environment. In the course of the environmental assessment, the share of mixed waste in general waste was also taken into account, which, on the one hand, illustrates the environmental self-discipline of residents (Hobson, 2020) and, on the other hand, demonstrates the effectiveness of city authorities in organizing processes within the framework of sustainable municipal waste management.

The above-described groups of conditions – defined for the purposes of statistical quantitative research – were then taken into account in the course of survey research of a qualitative, subjective nature, so that the evaluations obtained in the two research variants could be compared. This allowed for the confrontation of facts with opinions and made it possible to formulate more accurate conclusions and guidelines relating to the improvement of urban life. Combined with a review of international case studies, this enabled a multi-layered picture of how urban communities in Central and Eastern Europe function. The following section presents the layout of the survey questions used in the course of the qualitative research conducted in municipal offices.

The survey questionnaire used 40 determinants of residents' quality of life. They were rated by city officials responsible for developing and implementing city strategies on a five-point Likert scale. Their detailed list is included in Table 5.4.

Table 5.4 List of survey questions assigned to each area of development and sustainability of the surveyed cities

Area	Evaluated conditions
Economic (10 conditions)	Stability of the city's economic situation
	City's debt
	Overall investment attractiveness of the city
	Promotion of the city in terms of attracting new investors
	Possibility to get support for investment from the city funds
	Labor costs at the local level
	General employment opportunities in the city
	Loans for the creation of new jobs
	Financial support instruments for the unemployed
	Social assistance for the unemployed

(*Continued*)

Table 5.4 (Continued)

Area	Evaluated conditions
Technical and infrastructural (14 conditions)	Number of startups emerging in the city
	Support for SMEs in the application of smart solutions (innovation, new technologies, products and services)
	Level of functioning of entrepreneurship incubators in the city
	Level of functioning of technological parks in the city
	Availability of IT services
	Availability of R&D services
	Availability of housing in the city
	Newly built apartments in the city
	Availability of public transport
	Availability and density of road networks
	Availability and density of rail networks
	Availability and density of airline networks
	The municipal office has modern IT and office equipment (computers and office equipment)
	City Hall communicates with customers through social media
Socio-demographic (10 conditions)	Population growth
	Availability of primary education
	Availability of secondary education
	Availability of higher education
	Availability of training services
	Educational activities related to computer science beyond the curriculum, e.g., coding schools
	Percentage of university graduates in the population structure
	Sharing economy initiatives undertaken by the city (direct provision of services by people to each other, as well as sharing, co-creation, co-purchasing, etc.)
	Use of incentives for sharing economy organizations
	Existence of a person/cell in the city hall dealing with sharing economy issues
Environmental (6 conditions)	Relevance of environmental quality
	The difficulty of improving environmental quality
	The level of environmental pollution in the city compared to the national average in Poland
	Segregation of municipal waste in the city
	Percentage of residents using wastewater treatment plants compared to the national average
	Impact of environmental organizations on entrepreneurship in the city

Source: Own work.

In the case of **economic conditions**, the most important determinants of the city's development opportunities were considered to be factors directly related to the budget situation, which should be stable both in the short and long term. For these reasons, both the stability of the city's economic situation and its indebtedness were assessed, allowing to evaluate the current and future burdens of the city government and residents with liabilities that require repayment. In this regard, the assumption was made that a better financial situation and less debt are factors conducive to the development of smart cities and the improvement of the quality of life of their residents, since these circumstances are a determinant of overall urban prosperity. In addition, it is clear from the theoretical and review analyses carried out in the previous chapters that smart cities primarily include entities with very good financial standing, most often operating in highly developed economies.

In addition, the group of economic determinants also included factors defining the investment power of the city government. The overall investment attractiveness of the city, the promotional activities of the city government to attract new investors, as well as the possibility of obtaining financial support for new investments at the local level were evaluated. These are important developmental conditions, because smart cities are created and improved primarily based on the interaction of local governments with the business environment, which is the provider of modern information technology and telecommunications necessary to achieve Smart City status. Without the presence and interest of investors and their involvement in the construction of urban infrastructure, smart cities will not be able to develop. Local labor costs are also included in the group of investment determinants, as they are taken into account in the process of investor decision-making and are often a key determinant.

The third component of the questions on the economic determinants of smart city development was the problems of the local labor market, including the general possibilities of finding employment in the city and obtaining loans to create new jobs. Also, respondents were asked to assess the extent of financial and social support offered to the unemployed. The issues raised in this regard are important for the quality of life of residents dedicated to the idea of smart city development. Good employment conditions, low levels of unemployment and a wide range of support measures for the unemployed are factors that significantly determine the quality of urban life.

In the case of **technological and infrastructural determinants**, the first focus was on two key links of entrepreneurship and local innovation: startups and small and medium-sized enterprises, along with their potential and willingness to create innovation (Wolniak and Jonek-Kowalska, 2022). In this regard, the level of local, institutional support for new business initiatives in the form of the existence of business incubators and technology parks was also asked. This is because the literature suggests that city authorities and the extent of their innovation and entrepreneurship efforts can be an important pro-development factor.

In addition, this group of questions included issues related to the availability of IT and R&D services because of their direct connection with the development and implementation of smart city solutions in IT and ICT. The larger and more widespread it is, the faster the process of knowledge diffusion and completion becomes, which undoubtedly has a positive impact on the development of smart urban structures.

In the case of infrastructure considerations, the survey questionnaire first addressed the availability of housing and the extent to which the existing housing infrastructure is being injected with new resources. This is a particularly important issue for the quality of life of residents, who in Poland, as well as in other emerging and developing economies, face a shortage of vacant and affordable housing. A common problem in these economies is therefore the cohabitation of several generations, which certainly adversely affects the comfort of urban life. Hence, the described conditions were treated as a priority in the assessment of quality of life.

Transportation issues were further addressed by asking respondents to rate the availability of public transportation, as well as to assess the road, rail and air transportation networks. This is because in smart cities, transportation and logistics systems are being heavily developed and equipped with state-of-the-art operating and utilization technologies. So, their level of sophistication affects the maturity of a smart city and the efficiency, effectiveness and speed of movement of people and goods in the city, which in turn makes the residents' lives better.

The last two questions in the group of technical and infrastructural determinants concerned the modernity of the city government's operations, which manifests itself, among other things, in the use of advanced information and telecommunications technologies in the customer service process. Thus, the first question referred to the IT equipment at the city government's disposal, and the second to the use of social media as a communication channel. In the case of smart cities operating in highly developed economies, the issues indicated above seem obvious, but nevertheless in many smaller, less developed Polish cities – due to financial shortages – IT infrastructure does not always meet high international standards. Meanwhile, contact with local authorities and their information policies are important elements of building a city-community relationship (Yang and Lam, 2021). This relationship, in turn, is a prerequisite for creating higher-generation smart cities according to the idea of successive economic helixes.

In the group of **socio-demographic determinants**, the issue of population growth, which determines the reproduction of generations in urban communities, was placed first. Nowadays, especially in Europe, this is a very important issue due to the observed and systematically deepening aging of the population. This makes it necessary to reorient and adapt the Smart City concept to changing demographic factors. An example of this is the prioritization of goals related to senior care or health care, as well as the adaptation of the city's infrastructure to the needs of the elderly, disabled and sick. It is worth adding that this is a rather uncomfortable issue for smart cities due to its low spectacularity and incompatibility with the very exclusive image of the communities living in their territories. Despite the reality and perceptibility of the problem of aging urban communities, it is used relatively infrequently in the literature, but it will certainly gain importance in the future.

The questionnaire went on to include questions about the availability of education at the primary, secondary and higher levels, respectively, as well as questions related to the availability of training to improve professional skills. Issues related to the availability of lifelong learning options are an important development factor not only for smart cities, but also for entire economies (Bergman et al., 2018). After all, competent human resources, equipped with the latest knowledge, are the driving force for entrepreneurship and innovation, without which dynamic civilizational, social and economic development is not possible.

The final section of socio-demographic determinants asked about city government initiatives and involvement in the sharing economy. This is a form of sharing that characterizes the mature stages of smart city development, illustrating the maturity of the urban community in sustainable consumption ventures. Undoubtedly, it is not a priority in social development, but nevertheless it allows to assess to what extent the basic and own needs of residents have been met, so that they can think and focus on the common good and the need to save the planet. Sharing economy is therefore present in cities with high maturity in implementing smart city solutions.

The final section of the survey questionnaire includes questions relating to **environmental determinants** of urban quality of life (Liu and Zhang, 2021; Shamsuzzoha et al., 2021). This is a direct reference to the concept of sustainable development and the fivefold economic helix suggesting the inclusion of environmental organizations in the development of cities and regions. The first two questions addressed to respondents representing city offices referred to perceptions of environmental issues, which determine the scope and intensity of actions taken in this regard. Respondents were thus asked to determine how important environmental issues are to the city and how difficult it is to improve the city's environment. Subsequently, they were also asked about the level of environmental pollution in the city compared to the national average, so as to be able to confront the previously assessed views on environmental protection with the actual state of affairs in this regard.

The next two survey questions addressed specific determinants of environmental aspects of the city's quality of life. The first was related to an assessment of the extent of municipal waste segregation. The second was related to the percentage of residents using wastewater treatment plants compared to the national average in Poland. These are not only important determinants of attitudes to environmental protection but also elements to estimate the scale of residents' attitudes to environmental aspects. These are difficult issues due to the fact that smart cities are often associated with excessive consumerism, which is certainly not ecological. Also, residents themselves often do not understand the need for, and do not see the benefits of, environmental measures, as they are characterized by low environmental awareness.

Environmental organizations play a major role in its proper formation, which is why the last question in the survey questionnaire concerned the influence of environmental organizations on the city's business environment. Such participation is important not only from the point of view of the need to create sustainable cities but also, and perhaps especially, for modeling the desired environmental attitudes among businesses and residents.

The research procedure presented in this chapter was aimed at identifying the determinants of urban quality of life, taking into account contemporary guidelines for creating smart, responsible and sustainable cities. The following four chapters contain the results obtained using the proposed three-path approach for cities in Central and Eastern Europe, including Polish cities in particular.

Bibliography

Bank Danych Loklalnych. https://bdl.stat.gov.pl/bdl/start, [access data: 11.11.2022].

Batabyal, A., Nijkamp, P. (2019). Creative capital, information and communication technologies, and economic growth in smart cities. *Economics of Innovation and New Technology, 28*(2), 142–155. https://doi.org/10.1080/10438599.2018.1433587

Bergman, M., Ash, D., Osam, K., Strickler, B. (2018). Engineering the benefits of learning in the new learning economy. *The Journal of Continuing Higher Education, 66*(2), 67–76. https://doi.org/10.1080/07377363.2018.1469083

Bhattacharya, T.R., Bhattacharya, A., Mclellan, B., Tezuka, T. (2020). Sustainable smart city development framework for developing countries. *Urban Research & Practice, 13*(2), 180–212. https://doi.org/10.1080/17535069.2018.1537003

Chen, Ch.W. (2022). From smart cities to a happy and sustainable society: Urban happiness as a critical pathway toward sustainability transitions. *Local Environment.* https://doi.org/10.1080/13549839.2022.2119379.

Cugurullo, F. (2018). Exposing smart cities and eco-cities: Frankenstein urbanism and the sustainability challenges of the experimental city. *Environment and Planning A: Economy and Space, 50*(1), 73–92. https://doi.org/10.1177/0308518x17738535

Del-Real, C., Ward, Ch., Sartipi, M. (2021). What do people want in a smart city? Exploring the stakeholders' opinions, priorities and perceived barriers in a medium-sized city in the United States. *International Journal of Urban Sciences.* https://doi.org/10.1080/12265934.2021.1968939

Deng, D., Liu, S., Wallis, L., Duncan, E., McManus, P. (2017). Urban Sustainability Indicators: How do Australian city decision makers perceive and use global reporting standards? *Australian Geographer, 48*(3), 401–416. https://doi.org/10.1080/00049182.2016.1277074

Dohn, K., Kramarz, M., Przybylska, E. (2022). Interaction with city logistics stakeholders as a factor of the development of polish cities on the way to becoming smart cities. *Energies, 15*(11), 4103. https://doi.org/10.3390/en15114103

Hobson, K. (2020). From circular consumers to carriers of (unsustainable) practices: Socio-spatial transformations in the Circular City. *Urban Geography, 41*(6), 907–910. https://doi.org/10.1080/02723638.2020.1786329

IESE Cities in Motion Index. https://media.iese.edu/research/pdfs/ST-0542-E. pdf, [access data: 11.11.2022].

Jonek-Kowalska, I. (2022). Housing infrastructure as a determinant of quality of life in selected polish smart cities. *Smart Cities, 5*, 924–946. https://doi.org/10.3390/smartcities5030046

Kim, S.-C., Hong, P., Lee, T., Lee, A., Park, S.-H. (2022). Determining strategic priorities for smart city development: Case studies of South Korean and international smart cities. *Sustainability, 14*, 10001. https://doi.org/10.3390/su141610001

Komninos, N., Kakderi, C., Panori, A., Tsarchopoulos, P. (2019). Smart city planning from an evolutionary perspective. *Journal of Urban Technology, 26*(2), 3–20. https://doi.org/10.1080/10630732.2018.1485368

Liu, L., Zhang, Y. (2021). Smart environment design planning for smart city based on deep learning. *Sustainable Energy Technologies and Assessments, 47*, 101425. doi: 10.1016/j.seta.2021.101425.

Micozzi, N., Yigitcanlar, T. (2022). Understanding smart city policy: Insights from the strategy documents of 52 local governments. *Sustainability, 14*, 10164. https://doi.org/10.3390/su141610164

Monfaredzadeh, T., Berardi, U. (2015). Beneath the smart city: Dichotomy between sustainability and competitiveness. *International Journal of Sustainable Building Technology and Urban Development, 6*(3), 140–156. https://doi.org/10.1080/2093761X.2015.1057875

Mora, L., Deakin, M., Zhang, X., Batty, M., de Jong, M., Santi, P., Appio, F.P. (2021). Assembling sustainable smart city transitions: An interdisciplinary theoretical perspective. *Journal of Urban Technology, 28*(1–2), 1–27. https://doi.org/10.1080/10630732.2020.1834831

Orlowski, A. (2021). Smart cities concept – Readiness of city halls as a measure of reaching a smart city perception. *Cybernetics and Systems, 52*(5), 313–327. https://doi.org/10.1080/01969722.2020.1871224

Pineo, H., Zimmermann, N., Cosgrave, E., Aldridge, R.W., Acuto, M., Rutter, H. (2018). Promoting a healthy cities agenda through indicators: Development of a global urban environment and health index. *Cities & Health, 2*(1), 27–45. https://doi.org/10.1080/23748834.2018.1429180

PN-ISO 37120 Zrównoważony rozwój społeczny. Wskaźniki usług miejskich i jakości życia.

Shamsuzzoha, A., Nieminen, N., Piya, S., Rudledge, K. (2021). Smart city for sustainable environment: A comparison of participatory strategies from Helsinki, Singapore and London. *Cities, 114*, 103194. https://doi.org/10.1016/j.cities.2021.103194

Smętkowski, M., Moore-Cherry, N., Celińska-Janowicz, D. (2021). Spatial transformation, public policy and metropolitan governance: Secondary business districts in Dublin and Warsaw. *European Planning Studies, 29*(7), 1331–1352. https://doi.org/ 10.1080/09654313.2020.1856346

So, J., An, H., Lee, C. (2020). Defining smart mobility service levels via text mining. *Sustainability, 12*, 9293. https://doi.org/10.3390/su12219293

Su, Y., Fan, S. (2022). Smart cities and sustainable development. *Regional Studies*. https://doi.org/10.1080/00343404.2022.2106360

Su, Y., Hu, M., Yu, X. (2021). Does the development of smart cities help protect the environment? *Journal of Environmental Planning and Management*. https://doi.org/10.1080/09640568.2021.1999220

Suseno, Y., Salim, I.., Setiadi, P. (2017). Local contexts and organizational learning for innovation in an emerging economy: The case of two Malaysian firms in Indonesia. *Asia Pacific Business Review, 23*(4), 509–540. https://doi.org/10.1 080/13602381.2017.1346906

van den Buuse, D., van Winden, W., Schrama, W. (2021). Balancing exploration and exploitation in sustainable urban innovation: An ambidexterity perspective toward smart cities. *Journal of Urban Technology, 28*(1–2), 175–197. https:// doi.org/10.1080/10630732.2020.1835048

Wang, J., Liu, C., Zhou, L., Xu, J., Wang, J., Sang, Z. (2022). Progress of standardization of urban infrastructure in smart city. *Standards, 2*, 417–429. https:// doi.org/10.3390/standards2030028

Wawer, M., Grzesiuk, K., Jegorow, D. (2022). Smart mobility in a smart city in the context of generation Z sustainability, use of ICT, and participation. *Energies, 15*, 4651. https://doi.org/10.3390/en15134651

Wolniak, R., Jonek-Kowalska, I. (2021). The level of the quality of life in the city and its monitoring. *Innovation: The European Journal of Social Science Research, 34*(3), 376–398. https://doi.org/10.1080/13511610.2020.1828049

Wolniak, R., Jonek-Kowalska, I. (2022). The creative services sector in Polish cities. *Journal of Open Innovation: Technology, Market, and Complexity, 8*, 17. https://doi.org/10.3390/joitmc8010017

Yang, W., Lam, P.T.I. (2021). Evaluating non-market costs of ICT involving data transmission in smart cities. *Building Research & Information, 49*(7), 715–728. https://doi.org/10.1080/09613218.2020.1870426

6 Economic and financial determinants of the quality of life in Central and Eastern Europe cities

6.1 The economics of Smart Cities in Central and Eastern Europe

Economic and managerial factors are important determinants of the development of smart urban structures, as they directly determine the investment and technological capabilities of each city. Previous studies also indicate that urban structures are, by definition, privileged in terms of access to financing and the rate of economic growth. Kóňa et al. (2020), using Slovakia as an example, show that Slovak cities considered or aspiring to be smart are characterized by a very good economic situation compared to the country as a whole. Bratislava, in particular, stands out in this regard, but Košice, Banská Bystrica and Zvolen are not far behind. These entities are separated by quite a considerable distance from the economic indicators describing Slovakia's economic development. It is also notable that the country's capital, the aforementioned Bratislava, towers over the other cities equally in terms of innovation, leaving the other cities far behind. The described results allow us to conclude that smart cities – even in developing economies – are characterized by economic conditions that favor the implementation of smart urban solutions. Nevertheless, it should be emphasized that these conditions are more modest than in developed economies due to their lower level of economic development.

A broader study – conducted within the Visegrad Group (Poland, Hungary, Czech Republic, Slovakia) by Janusz and Kowalczyk (2022) – shows, in turn, that Czech cities (Prague, Brno, Ostrava and Plzeň) are best at implementing the Smart City concept, and cities in eastern Poland (Kielce, Rzeszów, Lublin) are weakest. The authors also argue that the Czech Republic's success is primarily due to the efficient implementation of e-government, significant economic and social potential, as well as the good state of the labor market, including openness to labor migrants. On the contrary, among the reasons for the failure of the cities of eastern Poland, they mention primarily the depopulation process in economically

DOI: 10.4324/9781003358190-6

less developed areas and the systematic aging of the population in this part of Poland.

Janusz and Kowalczyk (2022) – like the aforementioned Kóňa et al. (2020) – draw attention to the very large development gap between the capitals of the Visegrad Group countries and other cities, which should be bridged by directing financial support from the European Union to less developed regions, so that the Smart City concept has a chance to develop in a fully sustainable manner. The authors also identify key problems in the implementation of the Smart City concept, which are limited investment funds, lack of interest in long-term planning, centralization of decisions and lack of legislation supporting the development of smart city solutions.

The lack of a strategic approach to managing smart cities in developing economies is also demonstrated by Tantau and Santa (2021) in a comparison of SC development in Romania and Austria. The authors point out that Austrian cities have had and implemented holistic SC strategies for years, while in Romanian cities, the idea of SC is piecemeal and pilot in nature. For example, Vienna (the capital) of Austria has a strategy for Smart City development in the 2050 perspective.

Similar conclusions are reached by Naterer et al. (2018) who analyzed the strategies of Slovenian cities and related their content to the Europe 2020 Strategy. They noted that Slovenian documents are of poor quality and do not comply with European guidelines. This not only causes development and adaptation problems but also makes it difficult to set and control the directions of city authorities.

In addition, Varró and Szalai (2022) emphasize that in CEE cities, Smart Cities development patterns prevailing in developed economies cannot be uncritically adopted due to post-socialist institutional remnants. They also note that the basic priorities in the creation of smart city solutions in the analyzed region should be digitalization and economic and energy transformation. They consider European Union funding for infrastructure projects as an important determinant of the aforementioned processes.

In turn, research by Ban et al. (2022) shows that the economic problems associated with developing smart cities in developing economies are labor market constraints. In the case of Romanian metropolises (Oradea – analyzed by the aforementioned researchers), these are low wages and the lack of desirable jobs, especially those based on modern information and communication technologies.

However, the links between economic development and smart urban solutions are not everywhere confirmed. Indeed, a study by Popova and Popovs (2022) conducted in the Latvian economy shows that smart city economic growth directly affects only smart urban communities. The researchers did not confirm hypotheses indicating its impact on smart: living, mobility and environment. This allows an optimistic conjecture that smart city solutions can also be successfully created in developing economies.

Similar observations are made in the course of Senetra and Szarek-Iwaniuk's (2020) research. The authors note that small cities can also develop intensively and keep up with larger units, provided that they operate in a network in which knowledge and experience are exchanged. An example of such interaction is the Polish Cittaslow Network they analyzed, in which they manage to improve indicators of socioeconomic development despite differences in development potential.

The review shows that the development of smart cities in Central and Eastern Europe is possible and is taking place in many countries. Nevertheless, it is hampered by financial barriers. Management and organizational conditions, which are largely a legacy of the previous centrally planned economic system, are also an obstacle. It seems that financial problems can be offset by support for infrastructure projects from the European Union, while strategic and administrative difficulties should be worked on through the adaptation of good practices and consistent, long-term action.

6.2   Economy and finance in Polish Smart Cities – statistical perspective

This section presents an analysis of economic and financial conditions from a statistical perspective for 16 provincial cities. In this perspective, reference is made to three key indicators illustrating the situation of the surveyed units considered as:

- current in the context of budget income per capita (this is a key indicator of the wealth of local government units in Poland, it is used in comparative analyses and in the process of distributing subsidies and grants from the central budget; it reflects the current financial situation of the surveyed cities and the current possibilities for financing the development of smart urban solutions);
- strategic by looking at property expenditures per capita (this is a measure of an entity's commitment to urban infrastructure development that defines its investment capacity);
- debt illustrating the scale of the budget deficit/surplus in relation to total revenues (this is an indicator that, on the one hand, reflects the sufficiency of budget funds and, on the other hand, informs about the scale of debt accumulation in the case of budget deficits).

Thus, income per capita in 16 provincial cities in 2021 is shown in Figure 6.1. The highest value of the analyzed indicator characterized Warsaw

– the capital of Poland. In all other cities, it was below PLN 10,000. In addition to the capital city, the following units stood out from the rest in terms of income: Wrocław, Poznań and Opole with a budget income per capita exceeding PLN 9,000. The indicated units are large regional economic centers, which translates directly into their financial situation. They are also cities located in the western part of Poland considered to be a region better developed economically than the provinces located in the east of the country.

It is also worth noting that Wrocław and Warsaw – cities most often recognized as smart in international rankings (De Falco, 2019; Orlowski, 2021; Baran et al., 2022) – top the list. Thus, these are undoubtedly privileged entities in terms of wealth, which is certainly a convenience for them in the process of financing investments in the development of smart urban infrastructure. After all, as has been emphasized, the development of smart urban infrastructure requires significant capital investment (Biancardi et al., 2021).

The lowest value of budget income (less than PLN 7,000) was in Gorzów Wielkopolski and Toruń. Łódź and Białystok were also cities with fairly low budget income per capita. These are entities far less frequently mentioned in the rankings of smart cities, which do not have as much economic power as Warsaw or Wrocław. Less industrialization and population are associated in this case with lower budget revenues from income

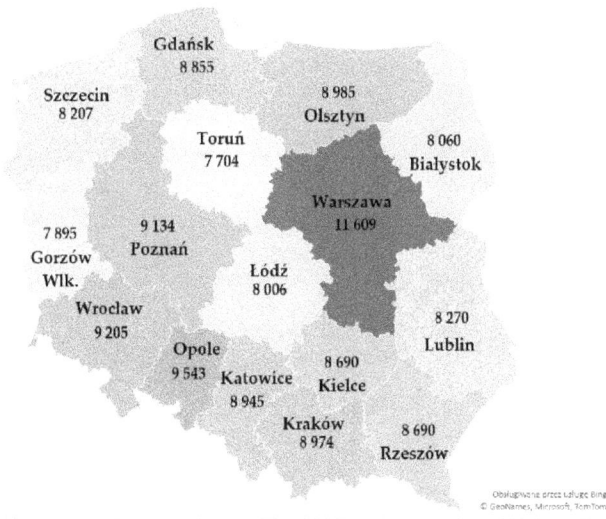

Figure 6.1 Budget revenue per capita in provincial cities in Poland in 2021 [in PLN].

Source: Own compilation using a map from Microsoft Excel based on information from the Local Data Bank.

and local taxes and can hinder the development of smart cities to a significant extent, since, as the previous chapters show, the creation of advanced technological and infrastructural solutions requires access to significant sources of financing (Blanck and Ribeiro, 2021; Jonek-Kowalska and Wolniak, 2021).

With the above statement in mind, Figure 6.2 shows the amount of property expenditures in the surveyed provincial cities in 2021.

The leaders in the above list were: Opole, Szczecin, Poznań and Gorzów Wielkopolski. Thus, the leaders included not only cities with a high degree of industrialization. Gorzów Wielkopolski, despite its low total budget revenues, devoted a significant part of them to property expenditures, which is a good indication of the strategic decisions of the city authorities, who are trying to improve the city's existing infrastructure.

Cities located in eastern Poland allocated considerably less funds for property expenditures. This region included two entities with the lowest level of the analyzed indicator (below PLN 1,000). These were Białystok and Olsztyn. This confirms the fairly common opinion about the division of Poland into region A (western, wealthier) and region B (eastern, poorer). It may also mean fewer opportunities for the development of smart urban solutions in economically less developed areas. It is interesting, however, that such regularity will not always occur, as evidenced by the example of Białystok, which is developing intensively and is sometimes

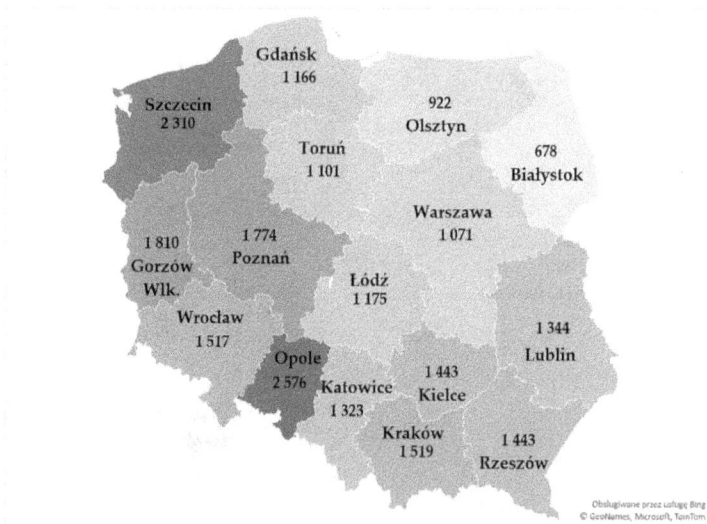

Figure 6.2 Property expenditures per capita in provincial cities in Poland in 2021 [in PLN].

Source: Own compilation using a map from Microsoft Excel based on information from the Local Data Bank.

mentioned in the literature and rankings as a contender for the title of fully smart city. Indeed, it should be remembered that the creation of smart city infrastructures also depends to a large extent on the efficiency and effectiveness of city management (Drapalova and Wegrich, 2020; Grossi et al., 2020; Lim et al., 2022; Park and Yoo, 2022), and favorable economic conditions are only one of the factors promoting their creation.

In the final stage of statistical analysis, reference was made to the sufficiency of budget resources, defined in relative terms as the ratio of deficit/surplus to total income. The value of this indicator for the surveyed provincial cities is shown in Figure 6.3.

According to the data presented in Figure 6.3, only 7 of the 16 analyzed cities had a budget deficit, with the highest level of this deficit in relation to total income recorded in Opole, Kraków, Szczecin and Gorzów Wielkopolski, which means that the entities listed above as leaders in property expenditures were implementing them at the expense of increasing the city's debt. This exposes the scope of Polish cities' financial problems.

Notably, many Polish cities have significantly increased the level of total debt in recent years as a result of the need to obtain an own contribution to finance infrastructure investments from European Union projects (mainly roads, rail routes, environmental protection) (Czykier-Wierzba, 2009; Mrozińska, 2017). This contribution was raised through

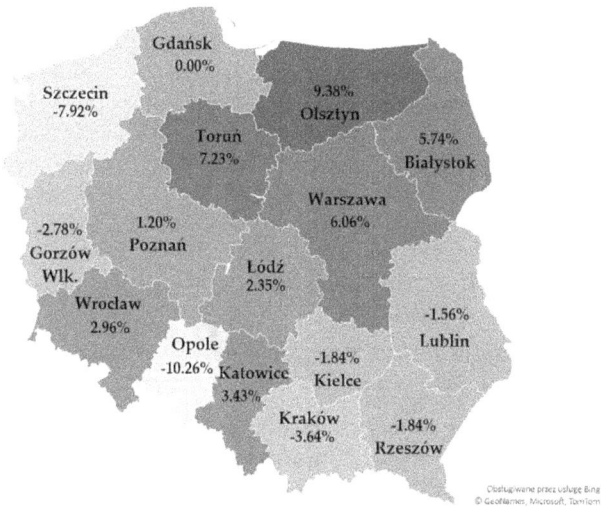

Figure 6.3 Budget deficit(−)/surplus(+) in relation to total budget revenues in provincial cities in Poland in 2021 [in %].

Source: Own compilation using a map from Microsoft Excel based on information from the Local Data Bank.

the issuance of municipal bonds and bank loans to the local government sector (Adamiak et al., 2012; Janeta, 2012; Hajdys, 2019). In many cases, the repayment period for this debt is estimated to be from a dozen to even several decades (Jonek-Kowalska and Turek, 2022). Hence, any additional budget deficit is a serious burden on the financial situation of Polish cities and may make it difficult to finance the property needs of future generations.

The map in Figure 6.3 also distinguishes four cities with the highest budget surplus (Olsztyn, Toruń, Warsaw and Białystok). Such a situation, of course, does not result in an increase in the city's debt and the need to seek deficit financing, but it is nevertheless difficult to assess it as correct, since public finance assumes annual budget balance (no surplus and no deficit). The high value of budget residuals may be an indication of poor budget planning or failure to carry out assumed tasks. This situation can be partially justified by the COVID-19 pandemic, but nevertheless its prolonged occurrence is not desirable.

In summary, the economic situation of Polish provincial cities is quite diverse. Large cities and those located in western Poland are characterized by better development conditions. Cities with a high level of property expenditures often finance infrastructure expansion from the budget deficit at the expense of increasing debt in the long term. In addition, the current and future economic situation (the COVID-19 pandemic; the Russia-Ukraine war, the conflict between national authorities and the European Union) is not conducive to the development of smart urban structures. Already the financial situation of local authorities and residents is deteriorating. Inflation is growing intensively, and unemployment is emerging. The economic recession is expected to deepen in 2023–2024. An important – for the community and the authorities – factor mitigating the effects of the economic crisis could be funds from the National Recovery and Resilience Plan obtained from the European Union (Dziembała and Kłos, 2021; Miernik, 2021), provided that the Polish government obtains an agreement with the European Commission, which is very difficult to obtain under the current political conditions and which further complicates the current economic situation.

6.3 Economic and financial conditions in Polish cities – survey perspective

In the next stage of the research, economic and financial considerations were presented from the perspective of municipal authorities based on surveys. A summary of the results including budget, investment and local labor market factors is included in Table 6.1.

Thus, on average, the stability of the city's economic situation was rated slightly higher than average by respondents, with 'good' being the

Table 6.1 Basic statistics for assessments of financial and economic conditions

Evaluated conditions	Arithmetic mean	Mode	Median	Standard deviation	Coefficient of variation
FINANCIAL					
Stability of the city's economic situation	3.68	4.00	4.00	0.89	24.34%
City's debt	3.14	3.00	3.00	1.03	32.76%
Overall investment attractiveness of the city	3.80	4.00	4.00	0.93	24.33%
Promotion of the city in terms of attracting new investors	3.71	4.00	4.00	1.01	27.22%
Possibility to get support for investment from the city funds	2.82	3.00	3.00	1.22	43.26%
ECONOMIC					
Labor costs at the local level	3.26	3.00	3.00	0.72	22.09%
General employment opportunities in the city	3.36	3.00	3.00	1.09	32.44%
Financial support instruments for the unemployed	3.22	3.00	3.00	1.04	32.30%
Loans for the creation of new jobs	2.48	3.00	3.00	1.29	52.02%
Social assistance for the unemployed	3.36	4.00	4.00	0.80	23.81%

Source: Own elaboration based on questionnaire surveys.

predominant rating, and responses between cities did not differ significantly. The surveys were conducted before the COVID-19 pandemic and the Russia-Ukraine war and therefore in good economic conditions during a phase of intensive economic growth (Wieczorek, 2017; Napiórkowski and Radło, 2022), which undoubtedly influenced respondents' final assessment of the cities' financial situation.

At this point, it is worth explaining that the budget of Polish municipalities is fed by their own revenues obtained from local fees and taxes (Dziemianowicz et al., 2018; Kossowski and Motek, 2021). Cities (urban municipalities) also share in income tax from individuals living in them and income tax from businesses operating on their territory. In addition, these units receive subsidies and grants from the central budget depending on the level of their own income – the lower it is, the higher the level of government support (Czempas, 2017; Głowicka-Wołoszyn et al., 2017).

It is clear from the above that the income of cities depends on economic development and industrialization and thus on local and central tax revenues (Jonek-Kowalska, 2018; Wichowska, 2021). For this reason, economic recession significantly affects the financial situation of Polish municipalities and poses a very serious risk to their development. Taking into account the rather difficult starting point of the surveyed cities related to functioning in a developing economy with relatively short free market traditions, the current crisis (Papińska-Kacperek, 2021) may threaten the development of smart city structures, since the Smart City concept is itself quite elitist, and it is difficult to think about its implementation under conditions of unsatisfied basic living needs of residents. Therefore, it can be expected that investments in this area will be put on hold for several years and possibly return to their continuation during the period of economic recovery (Kostyk-Siekierska, 2021).

Continuing with the description of the survey results, it can be noted that the assessment of the level of debt fell far short, which may imply an underestimation of the relationship between the expanding level of liabilities and the current budget situation and may be due to a tenure cognitive perspective that covers only the period until the next election. After all, as already mentioned, nationwide data show that the debt of the local government sector has been growing for many years, and in some cases, the repayment period of this debt reaches several decades (Wichowska, 2019; Wojtowicz and Hodzic, 2021; Białek-Jaworska, 2022), which undoubtedly poses a threat to current financial stability as well.

The evaluation discrepancies also relate to perceptions of the investment situation. Indeed, respondents gave a good rating for the city's overall investment situation. On the contrary, with regard to details, they were already more critical. They rated the city's promotional activities worse, and the possibility of obtaining investment support from city funds for investors was rated by them as unlikely.

Thus, based on the results obtained, it can be concluded that the overall economic and investment image of the city is quite good in the eyes of the authorities, but individual activities and their results no longer arouse so much enthusiasm. It is worth noting that such a duality of opinion may not serve as an objective assessment of the situation and may result in the buildup of economic problems and discouragement of investors from establishing operations in the city. This is because both entrepreneurs and residents evaluate the decisions and actions of the city authorities in the context of their detailed effects rather than intentions and generalities.

The ultimate quality of life is also affected by actual rather than potential management effects (Wołek and Hebel, 2019; Denis et al., 2021).

The assessment of the city's labor market from the city government's perspective is slightly better than average for both employers and employees. Indeed, respondents rated the level of labor costs in the city as average, as did employment opportunities. Financial support instruments for the unemployed, including social assistance for the unemployed, were also rated slightly better than adequate.

Interestingly, respondents rated very poorly the possibility of active support for the unemployed and investors in the form of loans for the creation of new jobs. This is an offshoot of an elaborate system of social assistance based on direct, not necessarily targeted, transfers of money to the unemployed and other social groups at risk of economic exclusion. Such a system of support – virtually devoid of incentive and activation instruments – can pose a serious threat to the social and economic development of cities. It also does not encourage entrepreneurship and innovation, which, like the circumstance indicated above, does not serve the development of smart cities.

Based on the above observations, one can conclude that Polish cities pay rather little attention to creating good conditions for investment. Meanwhile, this is the key to improving the local labor market and economic development. In the context of the budget principles described earlier, it is also a prerequisite for obtaining higher budget revenues from local and central taxes.

Poor economic development is particularly evident in cities where the restructuring of traditional industries has resulted in the liquidation of many large enterprises (mining, metallurgy, shipbuilding, etc.) and where attempts to build new economic structures have not been entirely successful. Such a situation is a derivative of the rather low innovativeness of the Polish economy and the enterprises operating in it. This makes it impossible to cope with modern technological challenges necessary to create advanced solutions in the area of information or telecommunications technologies. As a result, these cities are often doomed to depopulation and economic and social degradation (Szafrańska et al., 2019; Gołata and Kuropka, 2022). For these reasons, in this monograph the authors would like to draw attention to the relevance of knowledge and technology transfer, commercialization of scientific research results, innovation and entrepreneurship for the development not only of the economy as a whole but also of local and regional socioeconomic centers (Grimaldi et al., 2021; D'Amico et al., 2022).

In the course of further analysis, the survey results were contrasted with the size and wealth of the surveyed cities. An assessment of the relationships that occurred in this regard is included in Table 6.2.

The analysis found a fairly large number of statistically significant relationships between the studied determinants and the size and wealth of

Table 6.2 Basic statistics for assessments of financial and economic conditions

Evaluated conditions	Spearman's rank correlation coefficient	
	Population	Income level
FINANCIAL		
Stability of the city's economic situation	0.2339**	0.3987**
City's debt	0.0733	0.0987
Overall investment attractiveness of the city	0.3057**	0.3209**
Promotion of the city in terms of attracting new investors	0.3008**	0.2911**
Possibility to get support for investment from the city funds	0.3524**	0.3216**
ECONOMIC		
Labor costs at the local level	0.1617**	0.1817**
General employment opportunities in the city	0.3551**	0.3705**
Financial support instruments for the unemployed	0.3007**	0.2506**
Loans for the creation of new jobs	0.2307**	−0.1976**
Social assistance for the unemployed	0.1586**	0.0614

* $p < 0.05$; ** $p < 0.01$.

the surveyed cities. However, most of these relationships are of a weak or moderate nature, meaning that the number of residents and the level of income are related to the economic and financial situation of the city, but they are not relationships of a key nature.

Thus, the city's financial stability is most strongly correlated with the level of income, which seems quite obvious, since a higher level of income means spending sufficiency, greater development opportunities, and a lower risk of budget deficits and increased debt. So, cities with higher incomes, including those described in the first section, are more likely to develop and implement the infrastructural solutions of the Smart City concept.

According to respondents, higher incomes and a larger population also contribute to a more attractive investment image for the city. They also have a positive impact on the possibility of obtaining investment support from the city budget. It is worth adding, however, that while higher incomes directly affect the possibility of subsidizing city stakeholders, they do not necessarily determine the possibility of investment development, which also depends on the creativity and innovation of entrepreneurs or the local community. As an example, there are many smaller and less affluent localities that, taking advantage of their unique characteristics and resources, have thrived by creating new jobs and contributing to intensive local development (an example is the small town of Zator located in southern Poland, where one of the largest and most modern amusement parks in Europe has been created).

The responses also indicate that the size of the city and its wealth are related to promotion in terms of attracting new investors. It is difficult to agree completely with such a deterministic perception of this aspect. Undoubtedly, one hears about large and recognizable cities more often and more, but smaller regions also have their competitive advantages in the form of significant investment areas or lower costs of doing business. In addition, promotion, including its form, scope and strength of impact, depends directly on the initiative of the city authorities and local government actions taken in this regard. Adopting a passive stance in which failure is assumed in advance is certainly not appropriate and not conducive to development.

When analyzing the respondents' answers, it is also worth noting the lack of correlation between income and population and the level of city indebtedness. This confirms the previously described and widespread tendency for Polish cities to become indebted regardless of their size and wealth. Only the reasons for this may differ in this case. For smaller and poorer cities, rising indebtedness may be a result of a lack of funds to finance current budget expenditures. In large cities with higher incomes, it may result from financing property expenditures.

In assessing the economic factors determining residents' quality of life, the strongest correlations were in the impact of city size and wealth on employment opportunities. This is a reasonable observation due to the greater industrialization of large cities. These units are therefore able to offer their residents richer job opportunities and often higher wages.

Interestingly, the study shows that financial support options for the unemployed are more closely correlated with the size of a city than with its wealth as expressed by per capita income level. Nevertheless, as the population and income level increase, the unemployed can count on a greater range of financial support instruments. It is worth adding, however, that additional social assistance for the unemployed is weakly and insignificantly related to the size of the city and its wealth, respectively. This may mean that the already described rather wide range of social support is characterized by similar availability in both large and small cities.

A statistically significant but weak correlation relationship was found in the case of the dependence of the possibility of obtaining a loan for the creation of new jobs on the size and economic situation of the city, with a negative correlation in the case of the economic situation of the city, suggesting that the availability of this form of support is greater in smaller cities than in large entities. Entrepreneurs can therefore count on fairly even access to development funds, although unfortunately, from the assessment of this determinant described earlier, it is very limited. Newly emerging and expanding business entities must therefore rely primarily on their own ability to maintain business, which certainly does not encourage their creation and is not conducive to the development of smart urban structures.

The survey results presented in Table 6.2 also indicate a low correlation of local labor costs with the city's population and per capita income level. From the point of view of urban sustainability, this is good news, since the situation assessed in this way provides equal opportunities for all employees and employers regardless of the location and demographic and economic characteristics of the city. Nevertheless, combined with a low assessment of local labor costs – indicating their unattractiveness – this is not entirely positive information.

Consequently, based on the results of the surveys, we can conclude as follows:

- Of the financial considerations for living in the city, the possibility of obtaining investment support from the city's funds was rated the worst, and the promotion of the city in terms of attracting new investors was rated the best, although certainly the latter will not compensate for real help in establishing and developing a business.
- Among the economic determinants, the lowest rating was given to the possibility of obtaining a loan for the creation of new jobs, and the best rating was given to the general conditions of employment in the city and social assistance for the unemployed, which clearly indicates the passivity of forms of assistance for the unemployed and the low level of support for startups, especially small and medium-sized enterprises.
- There are many statistically significant relationships between city size and wealth and financial and economic determinants of quality of life, but they are weak and moderate in nature.
- The size of the city most strongly determines the possibility of receiving investment support from city funds and the overall employment opportunities in the city.
- The city's wealth is most strongly correlated with the stability of the city's economic situation and overall employment opportunities in the city.

Bibliography

Adamiak, J., Kołosowska, B., Voss, G. (2012). Obligacje komunalne na rynku Catalyst jako źródło finansowania działalności jednostek samorządu terytorialnego. *Annales Universitatis Mariae Curie-Skłodowska. Sectio H. Oeconomia, 46*(3), 101–109.

Ban, O.-I., Faur, M.-E., Botezat, E.-A., Ştefănescu, F., Gonczi, J. (2022). An IPA approach towards including citizens' perceptions into strategic decisions for smart cities in Romania. *Sustainability, 14*, 13294. htps://doi.org/10.3390/su142013294.

Baran, M., Kłos, M., Marchlewska-Patyk, K. (2022). Adaptacja miasta Warszawa do koncepcji smart city w oparciu o model odporności (resiliency model). *Przegląd Organizacji, 4*, 20–30. htps://doi.org/10.33141/po.2022.04.03

Białek-Jaworska, A. (2021). Revenue diversification and municipally owned companies' role in shaping the debt of municipalities. *Annals of Public and Cooperative Economic, 93*(4), 931–975. htps://doi.org/10.1111/apce.12358

Białek-Jaworska, A. (2022). Revenue diversification and municipally owned companies' role in shaping the debt of municipalities. *Annals of Public and Cooperative Economic, 93*, 931–975. htps://doi.org/10.1111/apce.12358

Biancardi, M., Di Bari, A., Villani, G. (2021). R&D investment decision on smart cities: Energy sustainability and opportunity. *Chaos, Solitons & Fractals, 153*, 111554. htps://doi.org/10.1016/j.chaos.2021.111554

Blanck, M., Ribeiro, J.L.D. (2021). Smart cities financing system: An empirical modelling from the European context. *Cities, 116*, 103268. htps://doi.org/10.1016/j.cities.2021.103268

Czepmas, J. (2017). Dysproporcje w dochodach gmin wiejskich w Polsce w latach 2002–2015. *Prace Naukowe Uniwersytetu Ekonomicznego we Wrocławiu, 485*, 59–68. htps://doi.org/10.15611/pn.2017.485.05

Czykier-Wierzba, D. (2009). Finansowanie z budżetu Unii Europejskiej zrównoważonego rozwoju miast w latach 2007–2013. *Zeszyty Naukowe Uniwersytetu Szczecińskiego. Ekonomiczne Problemy Usług, 37*, 9–12.

D'Amico, L., Boffa, D., Prencipe, A. (2022). University technology transfer and the contribution of university spin-offs in stimulating the socio-economic development of regions. *Advances in Management & Applied Economics, 12*(5), 19–33. htps://doi.org/10.47260/amae/1252

De Falco, S. (2019). Are smart cities global cities? A European perspective. *European Planning Studies, 27*(4), 759–783. htps://doi.org/10.1080/09654313.2019.1568396

Denis, M., Cysek-Pawlak, M.M., Krzysztofik, S., Majewska, A. (2021). Sustainable and vibrant cities. Opportunities and threats to the development of Polish cities. *Cities, 109*, 103014. htps://doi.org/10.1016/j.cities.2020.103014

Drapalova, E., Wegrich, K. (2020). Who governs 4.0? Varieties of smart cities. *Public Management Review, 22*(5), 668–686. htps://doi.org/10.1080/14719037.2020.1718191

Dziembała, M., Kłos, A. (2021). Pandemia COVID-19 a gospodarka Unii Europejskiej – instrumenty antykryzysowe oraz implikacje dla budżetu UE i jej państw członkowskich. *Przegląd Europejski, 1*, 81–98. htps://doi.org/10.31338/1641-2478pe.1.21.5

Dziemianowicz, R.I., Kargol-Wasiluk, A., Bołtromiuk, A. (2018). Samodzielność finansowa gmin w Polsce w kontekście koncepcji good governance. *Optimum. Economic Studies, 4*(94), 204–219. htps://doi.org/10.15290/oes.2018.04.94.16

Głowicka-Wołoszyn, R., Wołoszyn, A., Kozera, A. (2017). Nierówności dochodowe samorządów gminnych w Polsce. *Nierówności Społeczne a Wzrost Gospodarczy, 49*(1), 396–405. htps://doi.org/10.15584/nsawg.2017.1.30

Gołata, E., Kuropka, I. (2022). Large cities in Poland in face of demographic changes. *Bulletin of Geography. Socio-Economic Series, 34*, 17–31. htps://doi.org/10.1515/bog-2016-0032

Grimaldi, R., Kenney, M., Piccaluga, A. (2021). University technology transfer, regional specialization and local dynamics: Lessons from Italy. *The Journal of Technology Transfer, 46*, 855–865. htps://doi.org/10.1007/s10961-020-09804-7

Grossi, G., Meijer, A., Sargiacomo, M. (2020). A public management perspective on smart cities: 'Urban auditing' for management, governance and accountability. *Public Management Review, 22*(5), 633–647. htps://doi.org/10.1080/10.1080/14719037.2020.1733056

Hajdys, D. (2019). Architektura rynku obligacji komunalnych w Polsce w świetle obowiązujących przepisów prawa. *Finanse Komunalne, 3*, 16–32.

Janeta, A. (2012). Obligacje komunalne jako instrument finansowania rozwoju lokalnego i regionalnego. *Prace Naukowe Uniwersytetu Ekonomicznego we Wrocławiu, 271*(1), 236–246.

Janusz, M., Kowalczyk, M. (2022). How smart are V4 cities? Evidence from the multidimensional analysis. *Sustainability, 14*, 10313. htps://doi.org/10.3390/su141610313

Jonek-Kowalska, I. (2018). Kondycja finansowa jednostek samorządu terytorialnego jako determinanta rozwoju inteligentnych miast w Polsce. *Zeszyty Naukowe. Organizacja i Zarządzanie. Politechnika Śląska, 2018*, 131–140. htps://doi.org/10.29119/1641-3466.2018.120.10

Jonek-Kowalska, I., Turek, M. (2022). The economic situation of Polish cities in post-mining regions. Long-term analysis on the example of the Upper Silesian Coal Basin. *Energies, 15*(9), 3302. htps://doi.org/10.3390/en15093302

Jonek-Kowalska, I., Wolniak, R. (2021). Economic opportunities for creating smart cities in Poland. Does wealth matter? *Cities, 114*, 103222. htps://doi.org/10.1016/j.cities.2021.103222

Kóňa, A., Guťan, D., Horváth, P. (2020). Slovak republic on the way to build smart cities based on KPIs with first Slovak smart city index. *Scientific Papers of the University of Pardubice, Series D: Faculty of Economics and Administration, 28*(4), 1061.

Kossowski, T.M., Motek, P. (2021). Zróżnicowanie i polaryzacja przestrzenna dochodów własnych gmin. *Wiadomości Statystyczne, 66*(8), 1–23. htps://doi.org/10.5604/01.3001.0015.2301

Kostyk-Siekierska, K. (2021). Wpływ pandemii COVID-19 na sytuację finansową i funkcjonowanie jednostek samorządu terytorialnego. *Zeszyty Naukowe Małopolskiej Wyższej Szkoły Ekonomicznej w Tarnowie, 51*, 29–45.

Lim, Y., Edelenbos, J., Gianoli, A. (2022). Dynamics in the governance of smart cities: Insights from South Korean smart cities. *International Journal of Urban Sciences.* htps://doi.org/10.1080/12265934.2022.2063158

Miernik, R. (2021). Plan odbudowy dla Europy jako wzmocnienie bezpieczeństwa społecznego obywateli unii europejskiej po pandemii Covid-19. *Roczniki Nauk Społecznych, 14*(50), 193–205. htps://doi.org/10.18290/rns22501.7

Mrozińska, A. (2017). Struktura oraz skuteczność pozyskiwania środków na projekty unijne realizowane w miastach wojewódzkich w perspektywie 2007–2013. *Prace Naukowe Uniwersytetu Ekonomicznego We Wrocławiu, 467*, 57–69.

Napiórkowski, T.M., Radło, M. (2022). *Czynniki wzrostu gospodarczego regionów i podregionów województwa mazowieckiego.* Warszawa: Oficyna Wydawnicza SGH – Szkoła Główna Handlowa w Warszawie.

Naterer, A., Žižek, A., Lavrič, M. (2018). The quality of integrated urban strategies in light of the Europe 2020 strategy: The case of Slovenia. *Cities, 72,* 369–378. htps://doi.org/10.1016/j.cities.2017.09.016

Orlowski, A. (2021). Smart cities concept – Readiness of city halls as a measure of reaching a smart city perception. *Cybernetics and Systems, 52*(5), 313–327. htps://doi.org/10.1080/01969722.2020.1871224

Papińska-Kacperek, J. (2021). Zastosowanie idei smart city w czasie pandemii COVID-19 i izolacji społecznej w Polsce w 2020 roku. *Konwersatorium Wiedzy o Mieście, 34*(6), 63–72. htps://doi.org/10.18778/2543-9421.06.06

Park, J., Yoo, S. (2022). Evolution of the smart city: Three extensions to governance, sustainability, and decent urbanisation from an ICT-based urban solution. *International Journal of Urban Sciences.* htps://doi.org/10.1080/12265934 .2022.2110143

Parysek, J. (2017). Rewitalizacja miast w Polsce: wczoraj, dziś i być może jutro. *Studia Miejskie, 17,* 9–25.

Popova, Y., Popovs, S. (2022). Impact of smart economy on smart areas and mediation effect of national economy. *Sustainability, 14,* 2789. htps://doi. org/10.3390/su14052789

Senetra, A., Szarek-Iwaniuk, P. (2020). Socio-economic development of small towns in the Polish Cittaslow Network—A case study. *Cities, 103,* 102758. htps://doi.org/10.1016/j.cities.2020.102758

Szafrańska, E., Coudroy de Lille, L., Kazimierczak, J. (2019). Urban shrinkage and housing in a post-socialist city: Relationship between the demographic evolution and housing development in Łódź, Poland. *Journal of Housing and the Built Environment, 34,* 441–464. htps://doi.org/10.1007/s10901-018-9633-2

Tantau, A., Şanta, A.-M.I. (2021). New energy policy directions in the European Union developing the concept of smart cities. *Smart Cities, 4,* 241–252. htps://doi.org/10.3390/smartcities4010015

Varró, K., Szalai, A. (2022). Discourses and practices of the smart city in Central Eastern Europe: Insights from Hungary's 'big' cities. *Urban Research & Practice, 15*(5), 699–723. htps://doi.org/ 10.1080/17535069.2021. 1904276

Wichowska, A. (2019). Determinants of debt in rural municipalities on the example of the warmińsko-mazurskie voivodeship. *Acta Scientiarum Polonorum. Oeconomia, 18*(4), 121–130. htps://doi.org/10.22630/ASPE.2019.18.4.52

Wichowska, A. (2021). Determinanty dochodów własnych w budżetach kurczących się miast Polski w kontekście lokalnego rozwoju społeczno-gospodarczego. *Studia BAS, 4*(68), 27–45. htps://doi.org/10.31268/StudiaBAS.2021.34

Wieczorek, P. (2017). Makroekonomiczne uwarunkowania rozwoju gospodarczego –perspektywa dla Polski 2017–2019. *Kontrola Państwowa, 62/6*(377), 107–136.

Wojtowicz, K., Hodzic, S. (2021). Relationship between fiscal sustainability and efficiency: Evidence from large cities in Poland. *Economics and Sociology, 14*(3), 163–184. htps://doi.org/10.14254/2071-789X.2021/14-3/9

Wołek, M., Hebel, K. (2019). The quality of life in sustainable urban mobility planning. The case study of the Polish city of Piotrków Trybunalski. *Prace Naukowe Uniwersytetu Ekonomicznego We Wrocławiu, 63*(10), 129–143. htps:// doi.org/10.15611/pn.2019.10.10

7 Technological and infrastructure determinants of the quality of life in Central and Eastern Europe cities

7.1 Technological and infrastructure problems and challenges in selected cities of Central and Eastern Europe

An important factor in the development of smart cities is a high level of entrepreneurship, including the creation and functioning of startups. They are an incubator for new ideas and significantly strengthen innovation, which is one of the key determinants in the creation of smart urban solutions. Research on startups in smart cities in Central and Eastern Europe was conducted by, among others, Kézai et al. (2020). They show that in medium-sized cities in the Visegrad Group (Poland, Slovakia, Hungary, the Czech Republic), which aspire to be smart, the occurrence of startups is 19% higher than the national average. This means that these cities are able to offer potential entrepreneurs attractive business conditions, in return for which they can benefit from above-average economic growth. The conducted research also shows that startups develop best in: Brno in the Czech Republic, Bratislava in Slovakia, Debrecen and Szeged in Hungary, and in Bielsko-Biała, Gdańsk, Gliwice, Gdynia, Katowice, Poznań and Wrocław in Poland. A significant part of the identified entities operate in Poland, which is a distinction for this region of the Visegrad Group.

Research by Xydis et al. (2021) showed that in the case of the development of the technological infrastructure of smart cities, the most important and most often associated elements with the idea of Smart City are artificial intelligence and machine learning. The surveyed residents of Lithuanian, Danish, Slovak, Italian and Hungarian cities also positively perceived the implementation of smart energy distribution systems and the analysis of urban data. However, they expressed concerns about the use of smart sensors, smart cards and grade-smart meters. Reluctance to these solutions results from the attachment to the protection of privacy, as well as the fear of loss and undesirable use of personal data.

DOI: 10.4324/9781003358190-7

In addition to the technological solutions typical of the Smart City described above, residents also pay attention to the use of the assumptions of the Smart City concept in healthcare, which emphasizes another, non-technological determinant of the quality of urban life. It is shown in the surveys of Bălăşescu et al. (2022) conducted among residents of Romanian cities who believe that health should be the main priority when choosing directions for investing in modern urban technologies. This priority refers not only to raising healthcare standards but also to issues related to effective environmental protection that reduces climate and civilization threats to human health and life. This approach proves the high awareness of the surveyed inhabitants in terms of the sustainability of smart cities.

An important issue highlighted in the development of the concept of smart cities is also smart mobility. Examples of using solutions in this area become in practice a flagship achievement in terms of being smart. This is also typical of the countries of Central and Eastern Europe, which in recent years have received significant financial support from the European Union for the development of infrastructure: transport, road and railway. Thus, Ibănescu et al. (2022), examining Romanian cities aspiring to be Smart City state that most smart city solutions are implemented in the area of mobility. These are mainly Internet applications and platforms. The authors also point out the comprehensive monitoring of the implementation of the Smart City concept, so that there is no manipulation and abuse in this respect, which may happen in economies with relatively short free market traditions.

A study on mobility in Croatian smart cities conducted by Šurdonja et al. (2020) emphasizes the need to integrate technological and IT solutions with the city's social capital. The authors examine the attitude of residents to mobility solutions such as e-ticket, e-parking, transport adapted to demand, car sharing, bicycle sharing or information and mobile signaling. The obtained results indicate certain limitations in the acceptance of the above solutions by Croatian respondents, which implies the need to disseminate and educate the urban community in the field of benefits resulting from the development of smart cities.

In addition to transport and road infrastructure, housing infrastructure is an extremely important factor shaping the quality of life in cities. In the countries of Central and Eastern Europe, it is most often a remnant of socialist construction. It consists of large housing estates, which, despite the expiry of the assumed technical period of use, are still inhabited by successive generations. Nevertheless, according to research by Brade et al. (2009) they not only are a characteristic feature of the economies of this region but also enjoy widespread social acceptance, despite their unfavorable architectural image. The authors also focus on new urban trends, which are (1) the creation of new closed housing estates and (2) the construction of single-family houses in suburban areas. The latter solution is also described by Kubeš and Ouředníček (2022) in the example of Czech cities. They also emphasize that along with the creation of suburban housing estates of single-family houses outside the city, a commercial network located near local roads and highways is developing, so as to facilitate easy and quick shopping for residents. Notably, due to the low level of income of the majority of society in the economies of Central and Eastern Europe, these solutions are reserved for the wealthier, definitely less numerous part of society.

In the context of planning housing infrastructure, it should also be added that foreign investments also have a significant impact on its development, which is noted by Havel (2022) in the example of Poland. They contribute to the progressive neoliberalization of the real estate market and systematic departure from communist models. However, the observed changes lead to the minimization of the investment approval system in spatial planning, which speeds up administrative processes, but may also lead to abuses and disturbances in the urban housing infrastructure.

Brzeziński and Wyrwicka (2022) in the course of analyzing the implementation of the Smart City concept in Polish cities identify key development priorities that relate to sustainable housing infrastructure, smart mobility and smart energy. These are difficult and capital-intensive challenges, but without their implementation, not only will Poland not develop smart cities, but it will be difficult to think about dynamic economic development and catching up with Western European economies. Problems and state of implementation of the above-mentioned priorities are described in the next two sections.

7.2 Living and transport infrastructure in Smart City – statistical perspective

As emphasized in the theoretical part and in the previous section, one of the determinants of a smart city is technologically advanced transport infrastructure (Kramarz et al., 2021; Dohn et al., 2022). It is designed to facilitate the movement of people and goods and thus increase the comfort and efficiency of the urban transport network (Alnahari and Ariaratnam, 2022; Ghaffarpasand et al., 2022). It is worth emphasizing, however, that the development of the transport network requires significant capital expenditures due to the spatial scope and the advancement of modern urban transport infrastructure.

In the last 20 years, the transport infrastructure, including, above all, the network of local and regional roads and motorways, has developed significantly and improved quality thanks to the European Union funding (Pomykała and Engelhard, 2022). Nevertheless, there is still much to be done, especially in the area of implementing modern control technologies: lighting, public transport or the flow of goods (Pieriegud and Zawieska, 2018; Kachniewska, 2020).

In the comparative analysis of the surveyed provincial cities regarding transport infrastructure, data on the share of expenditure on transport and communications in total expenditure were used, which allowed to assess the scale of involvement of individual cities in improving this determinant of urban life. The results of this stage are presented in Figure 7.1.

Data presented in Figure 7.1 shows that the surveyed provincial cities allocate quite a significant part of the budget to transport and communications, ranging from about 14% in Katowice to over 26% in Poznań. This means that it is an important and significant development direction for the examined cities. The largest expenditures on transport and communications are incurred by Poznań, Szczecin, Warsaw and Lublin. These are economically well-developed cities, constituting centers of regional development, with a dense road and railway network. Their area also boasts airports. Białystok, Toruń, Katowice and Łódź spend the least funds on transport and communications, which in the case of the first two can be explained by the more agricultural character of the province. However, Łódź and Katowice are quite well-industrialized centers, and yet, the analyzed group of expenses is not as significant priority for them as for the leaders mentioned above.

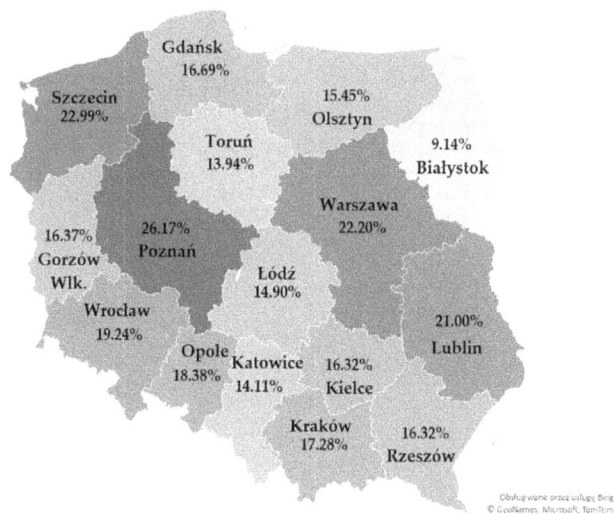

Figure 7.1 Share of expenditure on transport and communications in total budget expenditure in provincial cities in Poland in 2021 [in %].

Source: Own compilation using a map from Microsoft Excel based on information from the Local Data Bank.

In addition to transport and communication facilities, an important determinant of urban life is also the availability and use of housing infrastructure. Expenses for this purpose from the budgets of the analyzed municipalities are presented in Figure 7.2.

Data in Figure 7.2 and their comparison to the share of expenditure on transport and communications in the city budgets show that all cities spend significantly less budgetary resources on housing. The share of this type of expenditure ranges from 0.52% in Olsztyn to 8.38% in Szczecin. Notably, in 10 out of 16 provincial cities surveyed, it does not exceed 4%, which clearly indicates a lower rank of this expenditure priority.

The observed trends result from many years of problems and neglect of the housing economy in Poland, where municipal construction is practically not developed (Urban, 2020). The main reason for this is the lack of funds for such projects. This results in the systematic deterioration of the existing housing infrastructure and the lack of new housing units for the urban community (Nowak and Siatkowski, 2022). It is one of the most serious factors negatively affecting the quality of life in Polish cities, towns and villages (Rataj and Iwański, 2022). We should also add that the majority of government programs supporting housing construction are of a commercial nature, as they assume a significant share of own

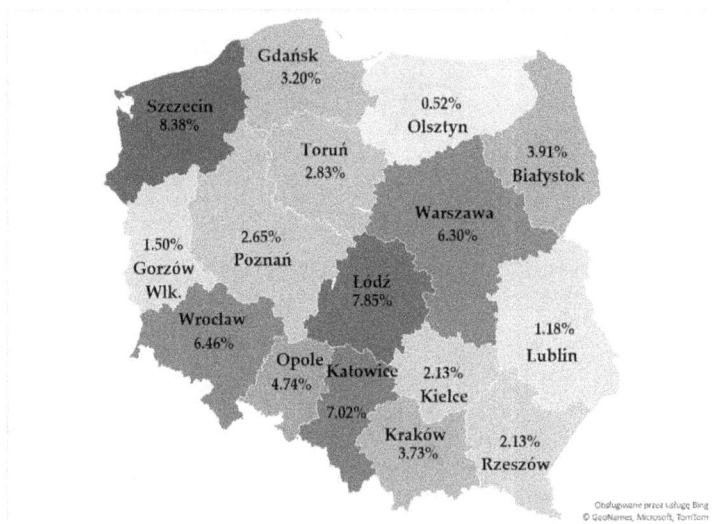

Figure 7.2 Share of expenditure on housing in total budget expenditure in provincial cities in Poland in 2021 [in %].

Source: Own compilation using a map from Microsoft Excel based on information from the Local Data Bank.

funds of potential apartment owners. This, in turn, limits their availability due to the low level of income and the lack or very low creditworthiness of most households in Poland (Czerniak et al., 2022). As a result, only people with above-average income can afford their own apartment or house. The rest of society uses properties built in the centrally planned economy, which are often inhabited by two or even three generations at the same time.

When reviewing urban infrastructural conditions, it can be stated that the examined cities devote much more attention and resources to transport and communications than to housing. As a result, the quality and accessibility of transport and communication infrastructure are better than housing infrastructure. At the same time, the development of the former has been supported for many years by funds from the European Union budget, which significantly influenced its current very good condition and scope. It should be added, however, that without proper attention to the availability of housing, it will be difficult to improve the quality of life in Polish cities. Research to date shows that the current availability of housing is one of the worst-rated determinants of life in Polish cities (Jonek-Kowalska, 2022).

7.3 Infrastructural and innovative conditions in Polish cities – survey perspective

This section describes the results of surveys relating to the technological and infrastructural determinants of the quality of life in Polish cities in the context of their aspirations to implement smart urban solutions. Basic descriptive statistics regarding the assessment of individual issues related to the above conditions are presented in Table 7.1.

Data contained in Table 7.1 show that among technological conditions, Polish cities rate access to IT services as the highest. Most of the respondents rated it as good; it also received the highest average rating. It can therefore be concluded that the digitization of the urban community is at a fairly good level, which generally creates favorable conditions for the implementation of smart urban solutions, including, for example, information or telecommunications applications.

Nevertheless, in the context of two subsequent assessments, it can be said that this access is passive and service-oriented, because the representatives of the city authorities find the availability of R&D services and the number of newly established startups in the city as much worse. Meanwhile, these are the key conditions necessary for the implementation and development of new technologies, so characteristic of smart cities (Biancardi et al., 2021; Clerici Maestosi, 2021). It can therefore be concluded that most Polish cities are characterized by a low level of entrepreneurship and innovation, which can be a significant obstacle in

Table 7.1 Basic statistics for assessing technological and infrastructural conditions

Evaluated conditions	Arithmetic mean	Mode	Median	Standard deviation	Coefficient of variation
TECHNOLOGICAL					
Number of startups emerging in the city	3.00	3.00	3.00	1.22	40.67%
Support for SMEs in the application of smart solutions (innovation, new technologies, products, services)	2.67	3.00	3.00	1.21	45.32%
Level of functioning of entrepreneurship incubators in the city	2.45	1.00	3.00	1.44	58.89%
Level of functioning of technological parks in the city	2.16	1.00	1.00	1.41	65.57%
Availability of IT services	3.47	4.00	4.00	1.02	29.39%
Availability of R&D services	3.04	3.00	3.00	1.21	39.80%
INFRASTRUCTURAL					
Availability of housing in the city	2.95	3.00	3.00	1.15	39.01%
Newly built apartments in the city	2.90	3.00	3.00	1.28	44.22%
Availability of public transport	3.36	4.00	4.00	1.14	33.93%
Availability and density of road networks	3.65	4.00	4.00	0.93	25.56%
Availability and density of rail networks	2.90	1.00	3.00	1.41	48.62%
Availability and density of airline networks	1.69	3.00	1.00	1.17	69.23%
City Hall has modern IT and office equipment (computers, office equipment).	3.93	4.00	4.00	0.81	20.61%
City Hall communicates with customers through social media	3.93	4.00	4.00	1.05	26.72%

Source: Own elaboration based on questionnaire surveys.

modernizing the infrastructure of Polish cities. Such observations are also confirmed by data and publications on the innovativeness of the Polish economy. In the European *Innovation Ranking Scoreboard 2021*, Poland as a country was in the group of the lowest rated countries – fourth from the end (Jankowska et al., 2022; Nawrocki and Jonek-Kowalska, 2022;

Tuznik and Jasinski, 2022). Previous research also shows that small and medium-sized enterprises are not interested in innovations and few of them undertake their development and implementation (Lewandowska and Cherniaiev, 2022; Zastempowski, 2022). The innovativeness of the Polish economy is also not conducive to the low pace of energy transformation, which could be a drive for pro-innovation activities on a large scale (Jonek-Kowalska, 2022; Rabiej-Sienicka et al., 2022).

The very low level of support in the use of smart solutions (innovations, new technologies, products, services) provided by the city authorities certainly does not encourage new, original solutions for small and medium-sized enterprises. Respondents give it an average rating of 2.67, which is worse than average. In such a situation, small and medium-sized enterprises are on their own. Therefore, they try to limit the risk to the necessary minimum, and as we know, innovations or startups are associated with above-average market risks, which increase the aversion to taking on challenges. The current post-pandemic economic crisis and the war in Ukraine will certainly not improve Polish innovation, because they mean the emergence of additional sources of threats of a fairly significant scale and intensity (Klein et al., 2022; Wojnicka-Sycz et al., 2022).

Among the examined conditions, the respondents assessed the lowest level of functioning of business incubators and technology parks in the city. This is due to the fact that they operate most often in large cities. This is also indicated by a large level of diversification of the respondents' answers. Nevertheless, in the light of the research results, the authorities of most Polish cities do not have the opportunity to use the results of R&D works, do not have funds to support entrepreneurship and innovation and cannot benefit from the activities of business incubators and technology parks, which significantly limits their technological capabilities, and thus constitutes a serious barrier to the design and implementation of smart urban solutions.

The infrastructural part of the survey opened with two questions about housing conditions related to the availability of apartments and newly constructed apartments in the city. In both cases, the dominant assessment of these conditions was 3.0, while the average indicated that both accessibility and new housing investments were rated below average. The obtained results confirm the housing problems of Polish cities described in the previous chapter. There is not enough housing, and unfortunately, not enough new and affordable housing infrastructure is being created. Such unfavorable conditions negatively affect the comfort and quality of life of residents and are directly felt by them.

In turn, the respondents positively assessed the issues related to the accessibility of road transport and the quality of road connections. Good marks prevailed in this respect, and the average was above average. This also confirms previous observations about the development of transport and road infrastructure in Polish cities financed by the European Union and quite significant budget expenditures.

Unfortunately, the accessibility and density of the railway network have not been assessed so well. Railway routes in Poland mainly connect large cities, which certainly hinders access to railway connections in small and medium-sized units. For many years, the development of railway routes has been accompanied by numerous technical, financial and organizational problems. Despite significant subsidies for railways from the European Union, the funds received are not fully used, which inhibits the improvement of the quality and accessibility of this means of transport (Bekisz et al., 2022; Korchagina et al., 2022).

Even worse scores were given by the respondents to air transport, because airports in Poland are located in 15 large cities. These are:

- Warsaw Chopin Airport (WAW);
- Bydgoszcz Ignacy Jan Paderewski Airport (BZG);
- Gdańsk-Rębiechowo Lech Wałęsa Airport (GDN);
- Poznań-Ławica Henryk Wieniawski Airport (PZN);
- John Paul II Kraków Airport (KRK);
- Katowice-Pyrzowice International Airport (KTW);
- Warsaw-Modlin Airport (WMI);
- Lublin Airport (LUZ);
- Wrocław Airport (WRO);
- Rzeszów-Jasionka Airport (RZE);
- Szczecin-Goleniów Airport (SZZ);
- Lodz Airport Central Poland (LCJ);
- Olsztyn-Mazury Airport (SZY);
- Zielona Góra-Babimost Airport (IEG);
- Warsaw-Radom Airport (RDO).

Considering the area and population of Poland, the location and number of airports may make it difficult to use airplanes as a means of transport. It should also be added that for many residents, air transport is not available or considered at all due to the costs.

In the context of the results obtained (for railways and air transport), the high level of diversification of the obtained ratings, which results from the distribution of railway stations and airports in large cities, should also be emphasized. Given the above, respondents from these cities more often gave higher marks in terms of the availability of rail and air connections.

Respondents significantly better surveyed the quality of the city's IT infrastructure. Both the modernity of the equipment and the ability to communicate with residents using social media (dominant) received a good rating and very high, above-average scores. This means that the vast majority of Polish municipal offices have the appropriate equipment to run e-services and digital communication. Therefore, they are quite well prepared for the implementation of smart city services and the development of the Smart City concept in this regard.

In the course of further considerations, reference was made to the relationship between the studied conditions and the size and wealth of the cities under study. Spearman's rank correlation coefficient illustrating the strength and direction of these relationships is presented in Table 7.2.

Data in Table 7.2 show that statistically significant relationships between the size of the city and the number of population were found for all technological conditions. These are positive correlations of weak or moderate strength. The strongest ones concern the functioning of business incubators and technology parks, which means that their presence is typical of large cities. A similar relationship was observed in the case of the number of newly established startups. The identified dependencies indicate that large cities are privileged in terms of the development of entrepreneurship and innovation.

Weaker correlations relate to support for SMEs in the use of smart solutions (innovations, new technologies, products, services); availability of IT services and availability of R&D services. Despite their low strength, it should be stated that they indicate a more favorable climate for the development of the Smart City concept in larger cities.

Table 7.2 Basic statistics for assessing technological and infrastructural conditions

Evaluated conditions	Spearman's rank correlation coefficient	
	Population	Income level
TECHNOLOGICAL		
Number of startups emerging in the city	0.4228**	0.2659**
Support for SMEs in the application of smart solutions (innovation, new technologies, products, services)	0.3802**	0.1848**
Level of functioning of entrepreneurship incubators in the city	0.4325**	0.1320*
Level of functioning of technological parks in the city	0.4390**	0.0955
Availability of IT services	0.3820**	0.1440*
Availability of R&D services	0.3676**	0.1099
INFRASTRUCTURAL		
Availability of housing in the city	0.4083**	0.3281**
Newly built apartments in the city	0.4287**	0.3109**
Availability of public transport	0.4229**	0.1348*
Availability and density of road networks	0.2889**	0.0932
Availability and density of rail networks	0.4257**	0.2005**
Availability and density of airline networks	0.2976**	0.1195*
City Hall has modern IT and office equipment (computers, office equipment).	0.1236	0.1640**
City Hall communicates with customers through social media	0.2336**	0.1742**

* $p < 0.05$; ** $p < 0.01$.

Moving on to the relationships between the level of income and the examined technological determinants of the quality of urban life, it can be seen that statistically significant but weak relationships relate primarily to the number of newly established startups and support for SMEs in the use of smart solutions. This is quite a valuable observation, because it indicates the need for the city to have good economic conditions for the development of innovative and small entrepreneurship.

Referring to the assessment of the relationship between infrastructure and the size and income of the city, it can be seen that, apart from modern IT equipment, all infrastructural conditions are positively and statistically significantly correlated with the size of the city. The strongest relationships concern the availability of existing and the scope of new housing infrastructure, which means that residents in large cities are offered better housing opportunities. Cities with a larger population are also characterized by greater availability of public transport and rail connections. The remaining relationships can be considered less significant due to the strength of the correlations found; however, they indicate a better quality of life in large cities resulting from better accessibility of housing and transport infrastructure.

A definitely smaller number of statistically significant relationships was found between the wealth of cities and the examined infrastructural conditions. The strongest of them concern housing infrastructure. This means that in cities with higher incomes, housing is more accessible, and investment in new housing infrastructure is also more frequent. This, in turn, has a very positive impact on the quality of life, as it is directly felt by the urban community. It also gives better chances for the sustainable development of smart cities.

Summarizing the considerations presented in this chapter, we can draw the following conclusions:

- The availability of IT services in cities is characterized by a good level, which may be a favorable condition for the development of the digital community and smart urban solutions.
- Entrepreneurship and innovation have low chances for development in Polish cities due to poor access to R&D services and the lack of startups, city support for small and medium-sized enterprises in the implementation of smart solutions, as well as very limited opportunities to use the services of business incubators and technology parks.
- The quality of life of the inhabitants of Polish cities is negatively affected by the lack of housing and the low level of development of new municipal construction.
- Significant expenditures on transport and connectivity supported by European Union funds allowed Polish cities to develop public transport and road networks well.

- Rail and air connections are located mainly in large cities, which results in their low accessibility in small and medium-sized cities and may negatively affect the quality of life of residents.
- The offices of Polish cities are well equipped with IT infrastructure, which should facilitate the process of digitization of public services and communication with city stakeholders and contribute to the use of the administrative assumptions of the Smart City concept.
- The level of entrepreneurship and innovation in the examined cities depends on their size expressed by the number of inhabitants and is definitely less dependent on the wealth of the city expressed by the level of income per capita.
- The development of housing infrastructure depends on the size and wealth of the city, which directly affects the quality of life of the inhabitants.
- The availability of public transport and rail connections is determined by the size of the city.

Bibliography

Alnahari, M.S., Ariaratnam, S.T. (2022). The application of blockchain technology to smart city infrastructure. *Smart Cities, 5*, 979–993. https://doi.org/10.3390/smartcities5030049

Bălășescu, S., Neacșu, N.A., Madar, A., Zamfirache, A., Bălășescu, M. (2022). Research of the smart city concept in Romanian cities. *Sustainability, 14*, 10004. https://doi.org/10.3390/su141610004

Bekisz, A., Kowacka, M., Kruszyński, M., Dudziak-Gajowiak, D., Debita, G. (2022). Risk management using network thinking methodology on the example of rail transport. *Energies, 15*, 5100. https://doi.org/10.3390/en15145100

Biancardi, M., Di Bari, A., Villani, G. (2021). R&D investment decision on smart cities: Energy sustainability and opportunity. *Chaos, Solitons & Fractals, 153*, 111554. https://doi.org/10.1016/j.chaos.2021.111554

Brade, I., Herfert, G., Wiest, K. (2009). Recent trends and future prospects of socio-spatial differentiation in urban regions of Central and Eastern Europe: A lull before the storm? *Cities, 26*(5), 233–244. https://doi.org/10.1016/j.cities.2009.05.001

Brzeziński, Ł., Wyrwicka, M.K. (2022). Fundamental directions of the development of the smart cities concept and solutions in Poland. *Energies, 15*, 8213. https://doi.org/10.3390/en1521821

Clerici Maestosi, P. (2021). Smart cities and positive energy districts: Urban perspectives in 2020. *Energies, 14*, 2351. https://doi.org/10.3390/en14092351

Czerniak, A., Czaplicki, M., Mokrogulski, M., Niedziółka, P. (2022). Financial affordability of housing in CEE countries amid changes in monetary policy. *Report of the SGH Warsaw School of Economics and the Economic Forum 2022.* Warsaw School of Economics: Warsaw.

Dohn, K., Kramarz, M., Przybylska, E. (2022). Interaction with city logistics stakeholders as a factor of the development of Polish cities on the way to becoming smart cities. *Energies, 15*, 4103. https://doi.org/10.3390/en15114103

Ghaffarpasand, O., Burke, M., Osei, L.K., Ursell, H., Chapman, S., Pope, F.D. (2022). Vehicle telematics for safer, cleaner and more sustainable urban transport: A review. *Sustainability, 14,* 16386. https://doi.org/10.3390/su142416386

Havel, M.B. (2022). Neoliberalization of urban policy-making and planning in post-socialist Poland – A distinctive path from the perspective of varieties of capitalism. *Cities, 127,* 103766. https://doi.org/10.1016/j.cities.2022.103766

Ibănescu, B.-C., Pascariu, G.C., Bănică, A., Bejenaru, I. (2022). Smart city: A critical assessment of the concept and its implementation in Romanian urban strategies. *Journal of Urban Management, 11*(2), 246–255. https://doi.org/10.1016/j.jum.2022.05.003

Jankowska, B., Mińska-Struzik, E., Bartosik-Purgat, M., Götz, M., Olejnik, I. (2022). Industry 4.0 technologies adoption: Barriers and their impact on Polish companies' innovation performance. *European Planning Studies.* https://doi.org/10.1080/09654313.2022.2068347

Jonek-Kowalska, I. (2022). Housing infrastructure as a determinant of quality of life in selected Polish smart cities. *Smart Cities, 5,* 924–946. https://doi.org/10.3390/smartcities5030046

Jonek-Kowalska, I. (2022). Multi-criteria evaluation of the effectiveness of energy policy in Central and Eastern European countries in a long-term perspective. *Energy Strategy Reviews, 44,* 100973. https://doi.org/10.1016/j.esr.2022.100973

Kachniewska, M. (2020). Factors and barriers to the development of smart urban mobility – The perspective of Polish medium-sized cities. In: Ujwary-Gil, A., Gancarczyk, M. (eds) *New Challenges in Economic Policy, Business, and Management.* Institute of Economics, Polish Academy of Sciences: Warsaw, 57–83.

Kézai, P.K., Szabolcs, F., Mihály, L. (2020). Smart economy and startup enterprises in the Visegrád countries—A comparative analysis based on the Crunchbase database. *Smart Cities, 3*(4), 1477–1494. https://doi.org/10.3390/smartcities3040070

Klein, M., Gutowska, E., Gutowski, P. (2022). Innovations in the T&L (transport and logistics) sector during the COVID-19 pandemic in Sweden, Germany and Poland. *Sustainability, 14,* 3323. https://doi.org/10.3390/su14063323

Korchagina, E., Desfonteines, L., Ray, S., Strekalova, N. (2022). Digitalization of transport communications as a tool for improving the quality of life. In: Rodionov, D., Kudryavtseva, T., Skhvediani, A., Berawi, M.A. (eds) *Innovations in Digital Economy. SPBPU IDE 2021. Communications in Computer and Information Science,* vol. 1619. Springer: Cham. https://doi.org/10.1007/978-3-031-14985-6_2

Kramarz, M., Knop, L., Przybylska, E., Dohn, K. (2021). Stakeholders of the multimodal freight transport ecosystem in Polish–Czech–Slovak cross-border area. *Energies, 14,* 2242. https://doi.org/10.3390/en14082242

Kubeš, J., Ouředníček, M. (2022). Functional types of suburban settlements around two differently sized Czech cities. *Cities, 127,* 103742. https://doi.org/10.1016/j.cities.2022.103742

Lewandowska, A., Cherniaiev, H. (2022). R&D Cooperation and investments concerning sustainable business innovation: Empirical evidence from Polish SMEs. *Sustainability, 14,* 9851. https://doi.org/10.3390/su14169851

Nawrocki, T.L., Jonek-Kowalska, I. (2022). Is innovation a risky business? A comparative analysis in high-tech and traditional industries in Poland. *Journal of Open Innovation, Technology and Market Complexity, 8*, 155. https://doi.org/10.3390/joitmc8030155

Nowak, M., Siatkowski, A. (2022). The late modernist community in the late socialistic block of flats: The issue of urban neighbourhood vitality in Poland. *Journal of Housing and the Built Environment, 37*, 101–123. https://doi.org/10.1007/s10901-021-09844-x

Pieriegud, J., Zawieska, J. (2018). Mobility-as-a-service – Global trends and implementation potential in urban areas in Poland. *Transport Economics and Logistics, 79*, 39–51. https://doi.org/10.26881/etil.2018.79.03

Pomykała, A., Engelhardt, J. (2022). Concepts of construction of high-speed rail in Poland in context to the European high-speed rail networks. *Socio-Economic Planning Sciences*, 101421. https://doi.org/10.1016/j.seps.2022.101421

Rabiej-Sienicka, K., Rudek, T.J., Wagner, A. (2022). Let it flow, our energy or bright future: Sociotechnical imaginaries of energy transition in Poland. *Energy Research & Social Science, 89*, 102568. https://doi.org/10.1016/j.erss.2022.102568

Rataj, Z., Iwański, R. (2022). The role of housing policy in long-term care in Poland. *Housing Policy Debate, 32*(4–5), 789–801. https://doi.org/10.1080/10511482.2020.1825011

Šurdonja, S., Giuffrè, T., Deluka-Tibljaš, A. (2020). Smart mobility solutions – Necessary precondition for a well-functioning smart city. *Transportation Research Procedia, 45*, 604–611. https://doi.org/10.1016/j.trpro.2020.03.051

Tuznik, F., Jasinski, A.H. (2022). Government incentives to stimulate innovation and entrepreneurship in Poland. In: Abdellatif, M.M., Tran-Nam, B., Ranga, M., Hodžić, S. (eds) *Government Incentives for Innovation and Entrepreneurship. Innovation, Technology, and Knowledge Management*. Springer: Cham. https://doi.org/10.1007/978-3-031-10119-9_7

Urban, F. (2020). Postmodernism and socialist mass housing in Poland. *Planning Perspectives, 35*(1), 27–60. https://doi.org/10.1080/02665433.2019.1672208

Wojnicka-Sycz, E., Piróg, K., Tutaj, J., Walentynowicz, P., Sycz, P., TenBrink, C. (2022). From adjustment to structural changes – Innovation activity of enterprises in the time of COVID-19 pandemic. *Innovation: The European Journal of Social Science Research*. https://doi.org/10.1080/13511610.2022.2036951

Xydis, G., Pagliaricci, L., Paužaitė, Ž., Grinis, V., Sallai, G., Bakonyi, P., Vician, R. (2021). SMARTIES Project: The survey of needs for municipalities and trainers for smart cities. *Challenges, 12*, 13. https://doi.org/10.3390/challe12010013

Zastempowski, M. (2022). What shapes innovation capability in micro-enterprises? New-to-the-market product and process perspective. *Journal of Open Innovation, Technology and Market Complexity, 8*, 59. https://doi.org/10.3390/joitmc8010059

8 Sociodemographic and educational determinants of the quality of life in Central and Eastern Europe cities

8.1 Needs, problems and challenges of local communities in Smart Cities in Central and Eastern Europe

Social issues are an important aspect of smart city sustainability, as it is assumed that the urban community should actively participate in the creation and implementation of smart city solutions. This is a key prerequisite for successfully improving the quality of life in a city. The main research directions in this area include participation, organizational and social networking, sharing economy and exclusion issues (Kolotouchkina et al., 2022; Sugandha et al., 2022).

Zait (2017) points out that the literature is still dominated by a hard-technological approach to the Smart City idea, which can hinder the implementation of the concept due to its lack of understanding or inaccessibility. The researcher emphasizes that in order to successfully create smart city solutions, it is also necessary to have and use soft competencies, such as entrepreneurial culture, discursive culture, everyday culture and civic culture (Zait, 2017).

In contrast, research on the social aspects of smart city development conducted in Romania by Rădulescu et al. (2020) points to the great importance of networks and cluster linkages in the process of knowledge and technology transfer. The authors also state that the involvement and networking of the urban community reinforce and perpetuate the achievements associated with the implementation of Smart City concepts.

Romanian researchers of smart city solutions also raise the problem of exclusion and even digital divide that occurs both at the level of individual cities and also among the urban community. Baltac (2019) notes that without bridging the indicated gaps and equalizing digital competencies, full implementation of the Smart City concept will never be possible.

The disabled and seniors are particularly at risk of social exclusion in smart cities. This exclusion can be digital, economic or social. This problem is particularly crucial due to the progressively aging urban

DOI: 10.4324/9781003358190-8

communities. Jonek-Kowalska (2022), in a study of the aging process in Polish provincial cities, draws attention to the need to provide effective and extensive healthcare for seniors (Jonek-Kowalska, 2022). This is an issue that significantly affects their quality of life and the sense of comfort and security of those who care for them.

Interestingly, Rink et al. (2014) criticize Central and Eastern European cities for being too liberal in their approach to the transformation of post-socialist urban structures. The researchers analyze the socioeconomic situation of four cities that functioned in the past in the so-called Eastern Bloc: Bytom (Poland), Ostrava (Czech Republic), Timisoara (Romania) and Leipzig (Germany). Of the aforementioned entities, the German Leipzig developed best in the new economic realities, managing not only to function well economically but also to retain its residents and prevent the process of shrinking urban communities. The other cities surveyed focused primarily on obtaining EU funds and foreign investment and their pragmatic use. At the same time, aspects such as the attractiveness of housing and preventing the depopulation of urban space were neglected. The results of the presented research expose the relevance and value of social issues in shaping the quality of life of residents. They also prove the practical importance of sustainable development of smart cities.

One of the key social issues described in the literature is public participation, understood as the inclusion of local communities in the urban development process. A study conducted by Vitálišová et al. (2021) in Slovakia shows that the most effective forms of participation include public discussion, annual reporting to local communities on the scope of local development goals, allowing residents to make public, in-person comments on strategic documents and participation of community stakeholders in working groups developing specific solutions for the city. In the practice of Slovak municipalities, the first two of these solutions are used by only 18% of the surveyed units. Even fewer, because only 8%, allow residents to comment on strategic activities. Allowing residents to participate in working groups is used by 44%, which means that this is the most popular form of public participation. Nevertheless, it should be noted that it was not rated as highly as the 3 previously mentioned by experts. This may suggest the city authorities' fear of open and unrestricted criticism, which is justified, but does not allow for the full confrontation of city plans with the expectations of the urban community. This, however, is a prerequisite for the development of fully smart and sustainable cities.

Unfortunately, research by Klimovský et al. (2016) also shows that citizens in Central and Eastern European cities show little interest in a participatory model of urban governance. They are also reluctant to use technological solutions that they perceive as being above their needs and expectations. According to the researchers, this is a major impediment to the development of smart cities in the region.

Despite the above barriers, the idea of smart cities in Central and Eastern Europe is steadily developing. As Kola-Bezka et al. (2016) point out, it is well-known and systematically implemented. It may not yet have reached the maturity characteristic of developed economies, but it is also not completely unnoticed or ignored. The authors' research shows that the Smart City concept is most often manifested and activated in the offering of modern public services to citizens by municipal authorities (Kola-Bezka et al., 2016). This shows their commitment to the process of modernizing the city and improving the quality of urban life.

8.2 Demographic and educational determinants of development in Polish Smart Cities – statistical perspective

The analysis of demographic and educational conditions in Polish cities began with a statistical review relating to 16 provincial cities. In the course of the research, data on the burden of elderly people in Polish society and the net enrollment rate were used, treating the indicated variables as basic in assessing potential demographic and educational problems of Polish cities. The conclusion of the section also refers to the issue of the responsibility of the communities of Polish cities and the attitude of residents to participation in decisions related to investment and development of Polish cities.

Thus, Figure 8.1 shows the percentage of people aged 65 and older in the total population in Poland's provincial cities in 2021.

Data in Figure 8.1 show that in 14 of the 16 cities surveyed, senior citizens comprise more than 20% of the urban community. The only exceptions in this regard are Lublin and Białystok, where the percentage is about 18%. It is worth adding that the process of aging of urban communities in Poland is steadily increasing over time, which is confirmed by the data in Figure 8.2 identifying the change in the percentage of people aged 65 and over in the total population in provincial cities in Poland between 2010 and 2021. In Gorzów Wielkopolski, Olsztyn, Toruń, Poznań and Opole, the number of residents aged 65+ increased by more than 40% over 12 years. In the other places, the increase was slightly smaller, but – except in Warsaw – was still about a third.

The above observations allow us to conclude that currently and in the future, one of the key problems for Polish cities will be the adaptation of urban infrastructure and smart city solutions to the needs of the aging population (Kowalczyk-Anioł et al., 2021; Zwierzchowska et al., 2021). It is therefore necessary to take this social group into account in the planning and development of smart cities. This is especially important given the issues of economic or technological exclusion of this part of the population accentuated in the literature (Wiig, 2016; Reuter, 2019; Radchenko, 2020).

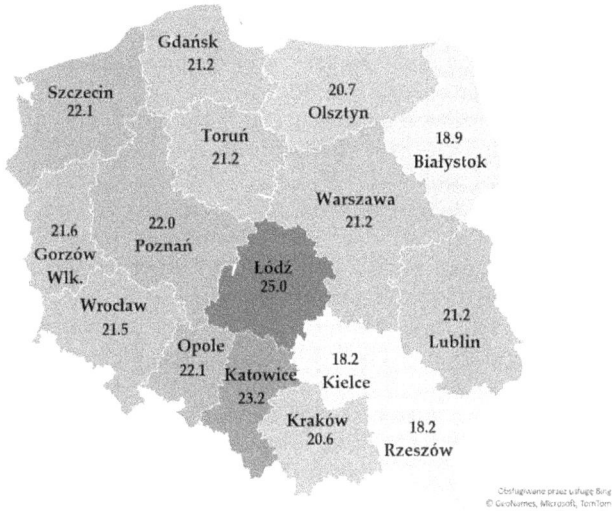

Figure 8.1 The percentage of people aged 65 and older in the total population in Poland's provincial cities in 2021 [in %].

Source: Own compilation using a map from Microsoft Excel based on information from the Local Data Bank.

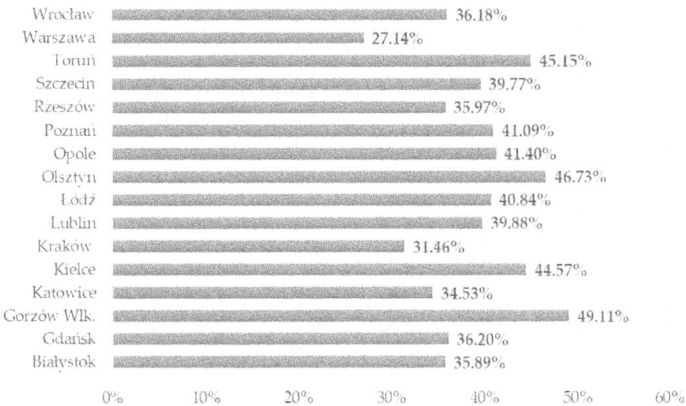

Figure 8.2 Change in the percentage of people aged 65 and older in the total population in Poland's provincial cities in 2010–2021 [in %].

Source: Own work.

The aging of urban communities is a derivative of very low birth rates (Cypryjański, 2019; Meardi and Guardiancich, 2022; Walaszek and Wilk, 2022). The value of this parameter for the provincial cities described in the chapter is shown in Figure 8.3.

Thus, according to the data in Figure 8.3, the demographic situation is worst in Łódź, Gorzów Wielkopolski, Katowice and Toruń. Most of these cities are also heavily burdened by a high percentage of seniors, further complicating their social conditions. The smallest gap between the number of live births and deaths, however, is found in cities such as Rzeszów, Kraków, Gdańsk, Wrocław and Warsaw. These are the entities most often mentioned in international rankings as smart. They are also distinguished by their very good visibility traits and economic situation. Thus, these cities attract people of working age, who, due to good living conditions, are more likely to make decisions to expand their families. The indicated circumstances allow these cities to achieve better results in terms of natural growth; nevertheless, it should be noted that there is no entity in the studied group that achieved positive natural growth in 2021, which illustrates the seriousness of the demographic situation of Poland and the surveyed cities.

The following analysis presents data on the net enrollment ratio for elementary schools (Figure 8.4). This coefficient determines the ratio of the number of people (in a given age group) studying (as of the beginning of the school year) at a given level of education to the population

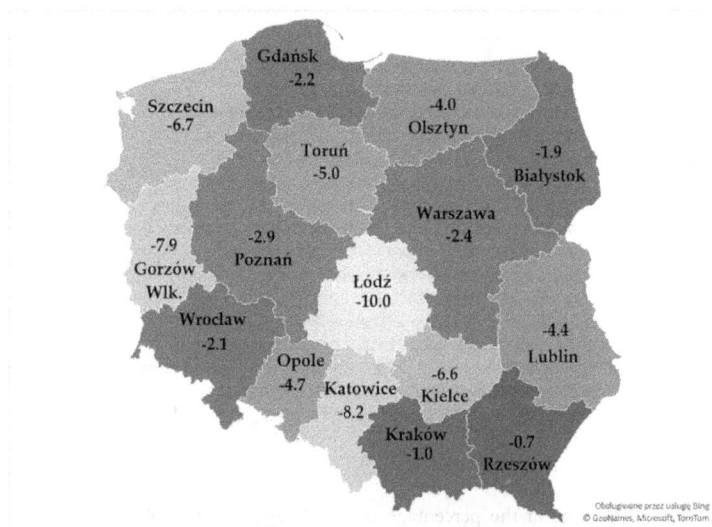

Figure 8.3 Population growth in provincial cities in Poland in 2021.

Source: Own compilation using a map from Microsoft Excel based on information from the Local Data Bank.

Figure 8.4 Net enrollment rate – elementary schools in provincial cities in Poland in 2021 [in %].

Source: Own compilation using a map from Microsoft Excel based on information from the Local Data Bank.

(as of December 31) in the age group defined as corresponding to that level of education. Thus, it allows us to answer the question of what proportion of children and adolescents benefit from the level of education intended for them.

The value of less than 100% occurs only in a few cities (Łódź, Gdańsk, Warsaw, Wrocław, and Gorzów Wielkopolski) and most likely results from population migration, which is a very good indication of the provision of basic education in Polish cities. This social need is therefore satisfied at a satisfactory civilization level. No irregularities and the existence of potential problems are identified.

The last described criterion for evaluating smart cities is social participation (Gao et al., 2020), understood in general as the will and active participation in decision-making about city development and participation in collective initiatives, of which sharing economy ventures can be a real dimension. The most common example of the use of public participation in Poland is civic budgets (Maziashvili and Kowalik, 2022; Miśkowiec and Masierek, 2022; Szczepańska et al., 2022), where city residents can decide on the selection of specified investments implemented in their district.

However, we should add that the very idea of public participation is not popular and still needs to be improved to match its level in highly developed countries. This is also confirmed by Poland's rather distant 51st place in the *Democracy Index 2021*. This ranking evaluates four main criteria: (1) electoral process and pluralism, (2) functioning of the government, (3) political participation and political culture and (4) civil liberties. Once the aggregate score is obtained, countries are divided into four groups: (1) full democracy, (2) flawed democracy, (3) hybrid regime and (4) authoritarian. Poland is in the latter group, which implies the need to improve democratic rules, including activating citizens for full participation in central and local strategic decisions.

Participation issues, however, are being recognized and cultivated. Participation is one of the ten thematic threads included in the *National Urban Policy 2023* adopted by the Polish government. In addition, in order to stimulate it, provisions have been introduced in the acts regulating the functioning of local governments to strengthen this aspect of urban life covering (*Social Participation. Report of the Ministry of Funds and Regional Policy*):

- the need to conduct public consultations based on the rules of the procedure previously adopted;
- the possibility of citizen resolution initiatives;
- coercion of the organization of the civic budget in cities with county rights;
- the openness of meetings of municipal councils (transmission, release of recordings, results of votes of councilors);
- ensuring residents' participation in the procedure for adopting or amending local spatial development plans and studies of land use conditions and directions;
- the obligation to maintain openness and transparency of planning procedures in the land use process;
- ensuring active participation of urban stakeholders at every stage of revitalization carried out in urban areas.

An assessment of the extent of social participation in Polish cities was undertaken by the Urban Policy Observatory (*Social Participation. Report of the Ministry of Funds and Regional Policy*). The research conducted by this center in 2018 shows that cities very often do not take active measures to increase public participation. Problems that are clearly noticeable in this regard are the lack of resolutions and civic consultations, inadequate standards of dialog, obstruction of the process of requesting public consultations and the lack of a clearer information policy to guarantee openness and transparency in decision-making processes undertaken by municipal authorities. Despite these impediments, residents declare an increasing willingness to participate and show trust in the city government (Radzik-Maruszak and Pawłowska, 2021; Pawłowska and Kołomycew, 2022). Moreover, opportunities for participation are growing faster than the willingness to use them. Thus, the observed trends allow a favorable prognosis for the development of various forms of public participation in the future.

8.3 Demography, education and participation in Polish cities – survey perspective

The next research stage refers to the results of surveys conducted in a group of Polish cities. Their results are presented in Table 8.1, along with basic statistics.

Data in Table 8.1 show that the majority of respondents have a very bad opinion of demographic growth in their city. This corresponds with the trends described in previous chapters indicating the systematic aging

Table 8.1 Basic statistics for assessments of demographic, educational and participatory determinants

Evaluated conditions	Arithmetic mean	Mode	Median	Standard deviation	Coefficient of variation
DEMOGRAPHICS					
Population growth	2.25	1.00	3.00	1.29	57.33%
EDUCATION					
Availability of primary education	4.22	4.00	4.00	0.70	16.59%
Availability of secondary education	3.82	4.00	4.00	1.06	27.75%
Availability of higher education	2.20	1.00	1.00	1.41	64.09%
Availability of training services	3.12	3.00	3.00	1.11	35.58%
Educational activities related to computer science beyond the curriculum, e.g., coding schools	2.89	3.00	3.00	1.25	43.25%
Percentage of university graduates in the population structure	3.00	3.00	3.00	1.11	37.00%
PARTICIPATION					
Sharing economy initiatives undertaken by the city (direct provision of services by people to each other, as well as sharing, co-creation, co-purchasing, etc.)	2.25	1.00	1.00	1.11	49.33%
Use of incentives for sharing economy organizations	1.97	1.00	1.00	1.17	59.39%
Existence of a person/ cell in the municipal office dealing with sharing economy issues	1.70	1.00	1.00	1.10	64.71%

Source: Own elaboration based on questionnaire surveys.

of Polish society. This is therefore noticeable and perceived also at the local government level. It is worth adding, however, that the rather high coefficient of variation means that not all cities are so critical of their demographic situation. The median value suggests that more than half assess the situation as better than 'bad' and 'very bad', but nevertheless this is not a satisfactory assessment, especially since a number of social and credit programs aimed at increasing fertility rates have been launched in Poland over the past ten years (Szczepaniak-Sienniak, 2021). Unfortunately, these have proven to be ineffective (Inglot, 2020), resulting in communities – even urban ones – aging at a very fast rate.

The evaluation of Polish cities in the educational area, especially in the area of primary and secondary education, is much better. Most respondents believe that the availability of primary and secondary schools offered by the city is at a good level. Nevertheless, a comparison of the median values for the variables described indicates that it is slightly worse in the case of high school education. The city government's assessment of primary education corresponds in this regard to the statistical evaluation carried out for Polish cities in the previous chapter on the basis of the enrollment rate.

As for universities, the evaluation of their accessibility is much lower (the predominant answer: 'very bad'). This is due to the fact that large universities, both private and public, operate primarily in large cities (most often in provincial capitals), which limits and hinders access to higher education for residents of smaller cities. This is also confirmed by the quite high value of the coefficient of variation illustrating the diversity of the surveyed units in this regard.

However, we should add that the percentage of people with higher education in the population structure is no longer rated as poorly as access to universities. The majority of respondents say that it is average, and moreover, this opinion does not vary significantly across the analyzed group. Therefore, it can be concluded that the residents, despite the potential difficulties in local access to universities, are mobile and successfully strive to raise the level of education. It should also be added that in the last 20 years, the percentage of people with higher education in Poland has increased significantly, which was also related to the establishment and development of many private universities as a result of the marketization of the Polish economy (Rybinski, 2020; Waligóra and Górski, 2021).

Respondents also rated on average the opportunities to improve their competencies through training, including IT training (although in the latter case, the rating is slightly worse). This could mean satisfactory prospects for the development of digitization and informatization of urban communities, so important for the implementation of the Smart City concept. However, it should be emphasized that the assessment in this regard is not sensational, especially since many social studies, including an overall assessment of the innovativeness of the Polish economy, show (Zygmunt, 2020) that Poles are not very keen on lifelong learning and are reluctant to take up additional forms of education (Kryk, 2016; Gnaj and Fijałkowska, 2021).

The last group of determinants assessed in the survey were social factors related to the development of the sharing economy, as one of the forms of social participation characteristic of the Smart City concept (Gori et al., 2015; Vinod Kumar and Dahiya, 2017). Surveys conducted in this area show that this is the most poorly perceived aspect of the operation of the surveyed cities (dominant and median of 1.00 – 'very bad' for all questions in this area). The idea of sharing economy is recognized by cities. They try to take initiatives in this area. Nevertheless, very few of them offer financial support for sharing ventures and have an institutionally separate position dealing with this area. The reason for this is certainly the insufficient level of financial resources and the focus of city authorities on the key livelihood needs of residents. A not insignificant factor may also be the relatively short period of operation of the Polish economy under free market conditions making it difficult to absorb patterns and behaviors more characteristic of emerging economies (Rutkowska-Gurak and Adamska, 2019; Jonek-Kowalska and Wolniak, 2022).

The following discussion refers to the analysis of the relationship between the population and financial situation of the cities under study and the demographic, social and educational factors analyzed. The results of the statistical analysis carried out in this regard are shown in Table 8.2.

Table 8.2 Basic statistics for assessments of demographic, educational and participatory determinants

Evaluated conditions	Spearman's rank correlation coefficient	
	Population	Income level
DEMOGRAPHICS		
Population growth	0.0901	0.0199
EDUCATION		
Availability of primary education	0.2838**	–0.0788
Availability of secondary education	0.5409**	0.1170
Availability of higher education	0.5312**	0.1923**
Availability of training services	0.4700**	0.2059**
Educational activities related to computer science beyond the curriculum, e.g., coding schools	0.2466**	0.2248**
Percentage of university graduates in the population structure	0.2496**	0.0866
PARTICIPATION		
Sharing economy initiatives undertaken by the city	0.3405**	0.1503
Use of incentives for sharing economy organizations	0.2831**	0.0825
Existence of a person/cell in the municipal office dealing with sharing economy issues	0.2933**	0.0399

* $p < 0.05$; ** $p < 0.01$.

Demographic determinants of natural increase do not depend on the size of the city and the level of budget income, which means that they are in no way related to the local or regional socioeconomic situation and are characteristic of all Polish cities. Low birth rates are therefore a nationwide trend.

In turn, all the educational determinants listed in the survey questionnaire are positively correlated with the size of the city expressed in terms of population, which means that as the size of the city increases, access to educational services increases. This relationship manifests itself most strongly in the case of secondary education, higher education and access to training. Thus, residents of large cities have greater access to the offer of competence improvement than those living in smaller units. Such a relationship indicates the educational privileging of large cities. To some extent, it also applies to cities with higher incomes, as evidenced by positive, statistically significant relationships between the availability of higher education, access to training services and IT-related educational activities beyond the curriculum and the city's financial health level.

The level of city income is not significantly correlated with the development of the sharing economy. In the group of participatory determinants, only statistically significant correlations were found between city size and initiatives and organizational and financial support for the sharing economy.

The correlation analysis shows that the level of demographic, social and educational determinants is more strongly related to the size of the city than to its wealth. Accordingly, it can be concluded that a large population promotes greater availability of services in the described field, which in turn makes it easier for large cities to strive for improved quality of life and Smart City status.

The key insights and observations on the educational, social and demographic conditions of Polish cities in light of the surveys allow the following conclusions:

- The only group of conditions rated at a good level (4.0) is access to primary and secondary education; all other factors scored below average (3.00).
- The authorities of the surveyed cities have the lowest opinion of demographic conditions (population growth) and sharing economy initiatives.
- Accordingly, most of the surveyed cities meet the analyzed groups of needs to a basic extent, far removed ideologically from the well-developed cities classified as Smart Cities.
- The surveyed cities are therefore threatened by the progressive aging of the population and a reduced interest in improving skills.

- The level of public participation is also low, despite legislative efforts to strengthen it and the willingness of residents to co-determine the directions of development of Polish cities.
- The size of the city has a low to average impact on educational determinants (availability of training and education at subsequent levels) and the level of development of the sharing economy.
- The city's wealth is positively, albeit weakly, correlated with the availability of higher education and training services.

Bibliography

Baltac, V. (2019). Smart cities—A view of societal aspects. *Smart Cities, 2,* 538–548. https://doi.org/10.3390/smartcities2040033

Cypryjański, J. (2019). Changes in seasonality of births in Poland in the years 1900–2009. *Demographic Research, 40,* 1441–54. https://www.jstor.org/stable/26727038

Gao, Z., Wang, S., Gu, J. (2020). Public participation in smart-city governance: A qualitative content analysis of public comments in urban China. *Sustainability, 12,* 8605. https://doi.org/10.3390/su12208605

Gnaj, I., Fijałkowska, B. (2021). Between a humanistic and economic model of life-long learning: The validation system in Poland. *European Journal of Education. Research, Development and Policy, 56*(3), 407–422. https://doi.org/10.1111/ejed.12466

Gori, P., Parcu, P.L., Stasi, M. (2015). *Smart cities and sharing economy.* Robert Schuman Centre for Advanced Studies Research Paper No. RSCAS 2015/96, Available at SSRN: https://ssrn.com/abstract=2706603 or https://doi.org/10.2139/ssrn.2706603

https://www.eiu.com/n/campaigns/democracy-index-2021/, [access data: 18.10.2022].

Inglot, T. (2020). The triumph of novelty over experience? Social policy responses to demographic crises in Hungary and Poland since EU enlargement. *East European Politics and Societies, 34*(4), 984–1004. https://doi.org/10.1177/0888325419874421

Jonek-Kowalska, I. (2022). Health care in cities perceived as smart in the context of population aging—A record from Poland. *Smart Cities, 5,* 1267–1292. https://doi.org/10.3390/smartcities5040065

Jonek-Kowalska, I., Wolniak, R. (2022). Sharing economies' initiatives in municipal authorities' perspective: Research evidence from Poland in the context of smart cities' development. *Sustainability, 14,* 2064. https://doi.org/10.3390/su14042064

Klimovský, D., Pinterič, U., Šaparnienė, D. (2016). Human limitations to introduction of smart cities: Comparative analysis from two CEE cities. *Transylvanian Review of Administrative Sciences,* 80–96. Available at: https://rtsa.ro/tras/index.php/tras/article/view/473

Kola-Bezka, M., Czupich, M., Ignasiak-Szulc, A. (2016). Smart cities in Central and Eastern Europe: Viable future or unfulfilled dream? *Journal of International Studies, 9*(1), 76–87. https://doi.org/10.14254/2071-8330.2016/9-1/6

Kolotouchkina, O., Barroso, C.L., Sánchez, J.L.M. (2022). Smart cities, the digital divide, and people with disabilities. *Cities, 123*, 103613. https://doi. org/10.1016/j.cities.2022.103613

Kowalczyk-Anioł, J., Łaszkiewicz, E., Warwas, I. (2021). Is the sharing economy inclusive? The age-related segmentation of Polish inhabitants from the perspective of the sharing economy in tourism. *Innovation: The European Journal of Social Science Research.* https://doi.org/10.1080/13511610.2021.1964347

Kryk, B. (2016). Accomplishment of the European Union lifelong learning objectives in Poland. *Oeconomia Copernicana, 7*(3), 389–404. https://doi. org/10.12775/OeC.2016.023

Maziashvili, M., Kowalik, I. (2022). City citizenship behavior and participation in promotion. *Place Branding and Public Diplomacy, 18*, 113–127. https://doi. org/10.1057/s41254-020-00194-z

Meardi, G., Guardiancich, I. (2022). Back to the familialist future: the rise of social policy for ruling populist radical right parties in Italy and Poland. *West European Politics, 45*(1), 129–153. https://doi.org/10.1080/01402382.2021.1916720

Miśkowiec, M., Masierek, E. (2022). Factors and levels of community participation using the example of small-scale regeneration interventions in selected neighbourhood spaces in Polish cities. *Urban Research & Practice.* https://doi. org/10.1080/17535069.2022.2099758

Partycypacja Społeczna. Raport Ministerstwa Funduszy i Polityki Regionalnej. http://irmir.pl/wp-content/uploads/2020/07/Partycypacja-spo%C5%82 eczna.pdf, [access data: 18.10.2022].

Pawłowska, A., Kołomycew, A. (2022). Local advisory councils in deliberative decision-making. Findings from research in Polish cities. *Journal of Contemporary European Studies, 30*(2), 345–362. https://doi.org/10.1080/14782804.2021.18731

Radchenko, K. (2020). From smart cities to smart regions: Regional economic specialization as a tool for development and inclusion. *Central and Eastern European EDem and EGov Days, 338*, 21–31. https://doi.org/10.24989/ocg.v.338.1.

Rădulescu, C.M., Slava, S., Rădulescu, A.T., Toader, R., Toader, D.-C., Boca, G.D. (2020). A pattern of collaborative networking for enhancing sustainability of smart cities. *Sustainability, 12*, 1042. https://doi.org/10.3390/su12031042

Radzik-Maruszak, K., Pawłowska, A. (2021). From advice and consultation to local cogovernance. The case of advisory councils in Polish cities. *Journal of Local Self-Government, 19*(4), 1043–1064.

Reuter, T.K. (2019). Human rights and the city: Including marginalized communities in urban development and smart cities. *Journal of Human Rights, 18*(4), 382–402. https://doi.org/10.1080/14754835.2019.1629887

Rink, D., Couch, Ch., Haase, A., Krzysztofik, R., Nadolu, B., Rumpel, P. (2014). The governance of urban shrinkage in cities of post-socialist Europe: Policies, strategies and actors. *Urban Research & Practice, 7*(3), 258–277. https://doi. org/10.1080/17535069.2014.966511

Rutkowska-Gurak, A., Adamska, A. (2019). Sharing economy and the city. *International Journal of Management and Economics, 55*(4), 346–368. https://doi. org/10.2478/ijme-2019-0026

Rybinski, K. (2020). Are rankings and accreditation related? Examining the dynamics of higher education in Poland. *Quality Assurance in Education, 28*(3), 193–204. https://doi.org/10.1108/QAE-03-2020-0032

Sugandha, Freestone, R., Favaro, P. (2022). The social sustainability of smart cities: A conceptual framework. *City, Culture and Society, 29*, 100460. https://doi.org/10.1016/j.ccs.2022.100460

Szczepaniak-Sienniak, J. (2021). Transformations of state family policy in Poland from 1989 to the pandemic period. *European Research Studies Journal, 24*(48), 883–900. https://doi.org/10.35808/ersj/2777

Szczepańska, A., Zagroba, M., Pietrzyk, K. (2022). Participatory budgeting as a method for improving public spaces in major Polish cities. *Social Indicators Research, 162*, 231–252. https://doi.org/10.1007/s11205-021-02831-3

Vinod Kumar, T.M., Dahiya, B. (2017). Smart economy in smart cities. In: Vinod Kumar, T. (eds) *Smart Economy in Smart Cities. Advances in 21st Century Human Settlements.* Springer: Singapore. https://doi.org/10.1007/978-981-10-1610-3_1

Vitálišová, K., Murray-Svidroňová, M., Jakuš-Muthová, N. (2021). Stakeholder participation in local governance as a key to local strategic development. *Cities, 118*, 103363. https://doi.org/10.1016/j.cities.2021.103363

Walaszek, M., Wilk, J. (2022). Population changes during the demographic transition. In: Churski, P., Kaczmarek, T. (eds) *Three Decades of Polish Socio-Economic Transformations. Economic Geography.* Springer: Cham. https://doi.org/10.1007/978-3-031-06108-0_10

Waligóra, A., Górski, M. (2021). Reform of higher education governance structures in Poland. *European Journal of Education. Research, Development and Policy, 57*(1), 21–32. https://doi.org/10.1111/ejed.12491

Wiig, A. (2016). The empty rhetoric of the smart city: From digital inclusion to economic promotion in Philadelphia. *Urban Geography, 37*(4), 535–553. https://doi.org/10.1080/02723638.2015.1065686

Zait, A. (2017). Exploring the role of civilizational competences for smart cities' development. *Transforming Government: People, Process and Policy, 11*(3), 377–392. https://doi.org/10.1108/TG-07-2016-0044

Zwierzchowska, I., Haase, D., Dushkova, D. (2021). Discovering the environmental potential of multi-family residential areas for nature-based solutions. A Central European cities perspective. *Landscape and Urban Planning, 206*, 103975. https://doi.org/10.1016/j.landurbplan.2020.103975

Zygmunt, A. (2020). Do human resources and the research system affect firms' innovation activities? Results from Poland and the Czech Republic. *Sustainability, 12*, 2519. https://doi.org/10.3390/su12062519

9 Environmental determinants of the quality of life in Central and Eastern Europe

9.1 Quality and environmental protection in Central and Eastern European cities

Environmental issues in emerging and developing countries are generally marginalized. This is primarily due to economic and political problems that determine the quality of life of the communities there. It seems that environmental awareness, and consequently recognition of the relevance of sustainability issues, emerges and grows after key livelihood needs have been met.

In the countries of the European Union, within the framework of solidarity mechanisms, efforts are being made to commoditize environmental policies and support the countries of Central and Eastern Europe in their efforts to implement the principles of sustainable development and corporate social responsibility. One of the tools used in this regard is the Cohesion Policy, which Serbanica and Constantin (2017) highlight in their research. Their analysis shows that the individual countries of the region focus on implementing smart pro-environmental solutions within the following specializations:

- eco-innovation and resource efficiency: mechatronics and green technologies (Bulgaria); ecological means of transportation (Czech Republic); use of natural resources in production and development of energy-efficient structures (Poland); new generation vehicles and resource recycling (Romania); eco-innovation in iron and steel production (Slovakia); modern heating systems; use of biomass and biomaterials; implementation of circular economy (Slovenia);
- smart, green and integrated transport: environmentally friendly means of transport (Poland); use of energy-efficient products for transport (Slovenia);
- sustainable agriculture: implementing innovations in agriculture, animal husbandry and hunting (Hungary);
- sustainable energy and renewable resources: use of new technologies in food production (Poland); clean and renewable energy (Hungary);

DOI: 10.4324/9781003358190-9

smart energy production and distribution systems (Lithuania and Lithuania); use of water resources (Romania); smart buildings and homes (Slovenia);

• sustainable water management and waste management: smart sewerage, water supply and waste management systems (Hungary); modern water renewal and water conservation technologies (Poland); smart use of resources to support the circular economy (Slovenia).

In protecting the environment in smart cities, the environmental awareness of residents plays a significant role. The greater it is, the easier it is to persuade them to act in support of sustainable consumption and production and climate protection. Unfortunately, this awareness is generally lower in developing economies than in developed economies, which is not conducive to the implementation of pro-environmental smart city solutions. For example, a study by Cepeliauskaite et al. (2021) comparing German cities with those in the Baltic Sea shows that in Kaunas (Lithuania), Riga (Latvia) and Tartu (Estonia), residents are very attached to private cars and are reluctant to use bicycles as a mode of transportation, even when the zero-carbon nature of this mode is emphasized. While in Berlin (Germany), the urban community is willing to give up the car in favor of bicycle commuting.

Perhaps the result of this approach to environmental protection is the rather low satisfaction of residents of CEE cities with climate quality. According to research conducted by Kopackova (2019) in Prague, Ostrava, Budapest, Miskolc and Warsaw, it appears that city authorities are barely or not at all involved and committed to climate issues. Surveyed residents rated the air quality and cleanliness of the cities in which they live very low as well.

Notably, interest in adapting city government policies to climate change is steadily growing in CEE countries, as noted by Kalbarczyk and Kalbarczyk (2022) in their analysis of 44 Polish cities. The authors conclude that institutional support is an important factor in the creation of pro-environmental plans and their implementation. They also emphasize that the entities they studied created very credible environmental plans, which were distinguished by a network approach to the implementation of goals and objectives, identified and fairly good climate change adaptation capabilities and consideration of uncertainties. Nevertheless, future assessments of the scale of implementation of the analyzed plans would be useful if only to compare declarations with implementation. This is especially important in the context of the previously cited survey results indicating poor performance and commitment of the city government to environmental and climate protection.

Improvements in the quality of urban sustainability plans are also noted by Erős et al. (2022) describing case studies from Romania. The authors find that the authorities of the cities studied are paying increasing attention to such issues as culture, society, health and public safety.

Unfortunately, Romanian cities still pay little attention to climate, the environment, innovation and waste management, pointing to the already stated marginalization of these issues in the CEE region. Nevertheless, without giving them a high enough priority, the development of smart and sustainable cities in the region will not be fully possible.

An earlier study by Lewandowska and Szymańska (2021) shows that greening in Polish cities is systematically progressing, but regrettably, at a slow pace. The authors examined 65 entities using an aggregate index that included 25 variables in the following areas: waste management, water and sewage management, air quality and protection, green infrastructure and nature conservation. The results they obtained show that over 13 years (the period 2004–2016), their proposed indicator increased by an average of 6%, which illustrates an improvement in environmental quality, but to a rather limited extent (below 0.5% per year).

Kronenberg et al. (2020) attempt to identify the weak interest and involvement of municipal authorities in environmental protection in postsocialist countries. Among the reasons, they point to tolerance of social inequality, lack of solidarity in society, lack of responsibility for the public interest, extreme individualization and disregard for social interests. According to the authors, the aforementioned reasons have led to the corporatization of city authorities and the formation of coalitions only between city authorities and the business environment, which is not conducive to sustainability and environmental justice.

From the above considerations, it is clear that the weakest side of urban development in CEE is the environmental side. Neither city authorities nor other stakeholders are interested in it, as they do not see the direct benefits associated with its quality and improvement. This is largely due to unmet basic livelihood needs and a Machiavellian perception of the necessary aspects of urban life.

9.2   Environment in Polish Smart Cities – statistical perspective

Protecting the environment in cities is quite a difficult challenge (Gherbi, 2012; Natanian and Auer, 2020). This is primarily due to the density of population in urban space and the strong industrialization of urban areas and their surroundings. Then, in a rather small area, there is an accumulation of problems related to the by-products of production and service processes and the functioning of households. Under such conditions, taking care of clean air, water and soil is very difficult and requires both a lot of money and very efficient management (Wang et al., 2015; Lin et al., 2021). For these reasons, this section examines selected environmental issues that are characteristic of Poland's provincial cities.

In the course of the research, the first reference was made to the general perspective expressed in terms of the level of environmental and climate protection expenditures per capita. However, before the final analysis of this parameter, let's refer to the distribution of environmental protection tasks in the Polish local government system, which will help to understand and explain the results obtained. According to the current *Act on Commune Self-government* (Journal of Laws 1990 No. 16 item 95, as amended), meeting the collective needs of the community is one of the municipality's own tasks. In particular, own tasks include matters of (…) environmental and nature protection and water management, (…) municipal greenery and tree planting. The general policy of shaping the environment following the principle of sustainable development is, in turn, carried out at the provincial level in accordance with the *Act on Voivodship Government* (Journal of Laws 1998 No. 91 item 576, as amended). The provincial government also performs environmental protection tasks at the level of the entire region. Therefore, in the context of the above regulations, it can be concluded that the environmental protection strategy is a regional task, while the real activities of its implementation lie with the municipalities, including the cities analyzed in this study (Witkowski, 2019; Kowalczyk, 2022; Satoła and Milewska, 2022).

Figure 9.1 thus shows how much the surveyed cities spent on environmental protection in 2021. For comparative purposes, the level of actual spending is presented on a per capita basis.

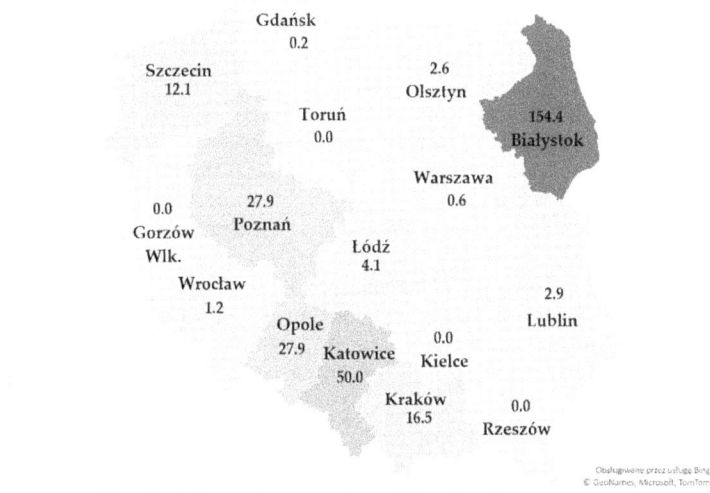

Figure 9.1 Air and environmental protection expenditures per capita in provincial cities in Poland in 2021 [in PLN].

Source: Own compilation using a map from Microsoft Excel based on information from the Local Data Bank.

According to the data in Figure 9.1, the level of environmental and climate protection expenditures in 2021 varied widely among the surveyed cities. Cities such as Toruń, Gorzów Wielkopolski, Rzeszów and Kielce did not incur them at all. In many other entities, these expenditures did not exceed PLN 10, which should be considered a symbolic sum, even on a per capita basis. Slightly higher involvement in this regard (from PLN 11 to PLN 30) was recorded in Szczecin, Poznań, Opole and Kraków. In general, these are fairly well-recognized cities, most of them highly industrialized, which perhaps provides them with an additional incentive to take action on environmental protection. A similar rationale may accompany Katowice, which spent as much as PLN 50 on environmental and climate protection. We should emphasize, however, this is a post-mining area, heavily exposed to the effects of years of mining, which may be related to both the need and greater availability of funds for environmental and climate protection. The largest environmental expenditures in 2021 were noted in Białystok, which is particularly noteworthy because it is a city located in a very green region of Poland, and yet, it attaches great importance to environmental issues. Significant spending in this area was related to the pro-environmental reconstruction of the Biała River bed. Notably, the city devotes a lot of effort to environmental education and activities on its website (https://www.bialystok.pl), which certainly promotes the sustainable development of the place.

In the context of the above observations, it can be concluded that in most of the surveyed cities, little attention is paid to environmental and climate issues, reflecting a certain marginalization of these problems. The existing situation is certainly related to the post-pandemic crisis and the inadequacy of budget resources, but it nevertheless reveals the very weak position of environmental priorities in local municipal budgets. This can be considered quite typical of developing economies (Chodkowska-Miszczuk et al., 2021; Mesjasz-Lech, 2021; Zawilińska, 2021), where infrastructure or social assistance needs are more pressing. Nevertheless, adopting such a perspective poses a threat to the quality of life of future generations.

An important outcome of the environmental measures of smart cities, often presented in the literature, is the green infrastructure created as part of the Smart City concept and serving residents for recreational purposes and supporting the process of cleaning the city's air (Anguluri and Narayanan, 2017; Artmann et al., 2019; Kaluarachchi, 2021). For these reasons, Figure 9.2 presents the share of parks, greens and neighborhood green areas in total area in provincial cities in Poland in 2021. Data presented in Figure 9.2 show that this share ranged from 1.5% (Gdańsk) to 13.3% (Kielce and Rzeszów). Moreover, it is significantly higher in cities located in eastern and central Poland than in western cities. Gdańsk and Szczecin, port cities located in the Baltic Sea region, had the lowest rates in this regard.

Generally speaking, there is more agricultural and tourist land in the eastern part of Poland, which could favor their natural development by cities there into parks or green spaces. Nevertheless, it should be noted that these cities, due to favorable environmental conditions, could abandon their creation (due to natural access to green areas); however, they are actively working to create them. Notably, it is Wrocław that stands out against western cities, with a large share of parks and green spaces for this part of Poland (Niedźwiecka-Filipiak et al., 2015; Niedźwiecka-Filipiak et al., 2019), which confirms the validity of its position in international rankings of smart cities.

The above-cited comparative summary illustrated in Figure 9.2 is worth contrasting with the share of green areas in total area in provincial cities in Poland in 2021. It reflects the total green areas located in the surveyed cities, and therefore, both those created by the city and those that are the work of nature and left in the city in a significantly unchanged condition. The share of green areas in the total area is shown in Figure 9.3.

Katowice, as a highly industrialized and urbanized city, fared significantly worse in the modified ranking. The very low rating of Gorzów Wielkopolski and Opole did not improve significantly either. Rzeszów and Kielce, on the one hand, were still among the leaders in creating

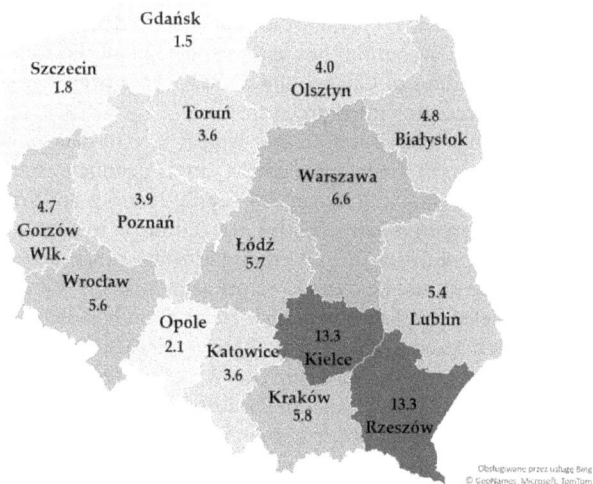

Figure 9.2 Share of parks, greens and neighborhood green areas in total area in provincial cities in Poland in 2021 [in %].

Source: Own work using a map from Microsoft Excel based on information from the Local Data Bank.

Figure 9.3 Share of green areas in total area in provincial cities in Poland in 2021 [in %].

Source: Own work using a map from Microsoft Excel based on information from the Local Data Bank.

green areas. Olsztyn, on the other hand, moved to the forefront, with urban greenery accounting for as much as 22% of the total area. Interestingly, the rating of Gdańsk and Szczecin improved significantly after taking into account natural green spaces.

Given the results, however, one would have to conclude that a 5 or even 10% share of green space in the city is definitely insufficient, as it means that the city is dominated by dense housing and road construction, which prevents any effective recreation or leisure activities. Certainly, such a situation negatively impacts the urban ecosphere. Thus, the following should be considered primarily smart green cities: Olsztyn, Poznań, Kielce and Rzeszów, which are undoubtedly helped in the creation of green spaces by their natural geographical location. Furthermore, it would be appropriate to single out Wrocław, Łódź, Kraków and Warsaw, which, as more populous and industrialized cities, strive to maintain more than a 10% share of green areas in the total city area.

The last environmental indicator addressed is municipal waste management in the surveyed cities (Hasan, 2004). This is an important issue given that smart cities are often accused of excessive consumerism (Chekima et al., 2016; Never and Albert, 2021), which results in above-average consumption of natural resources and the generation of excessive pollution. For these reasons, Figure 9.4 shows the share of mixed waste in total waste in the surveyed cities in 2021, allowing us to assess the scale of segregation and thus the city government's efforts toward sustainable waste management.

Data in Figure 9.4 show that in all cities studied, mixed and therefore unsegregated, waste exceeded 50% of the total waste generated by households. The largest amount of waste was not segregated in Katowice, Olsztyn, Szczecin, Toruń and Warsaw. Rzeszów, Białystok and Kraków fared best on the list. Notably, in Polish cities most often considered smart, waste segregation was just average (Warsaw – 66.64% and Wrocław – 63.46% of mixed waste). Thus, the results obtained in the analyzed scope testify to the not yet fully utilized opportunities for smart municipal waste management.

Summarizing the considerations undertaken in this section, it should be stated that the most favorable environmental conditions characterize Białystok, where the highest amount of budget expenditures per capita is allocated to environmental and climate protection, the share of green areas in the total area is above average and the volume of segregated waste is the highest. In contrast, the worst environmental situation is in Gorzów Wielkopolski, which lacks spending on environmental and climate protection, there are few green areas and the share of mixed waste in total waste is quite high. Also, Warsaw and Wrocław look quite average in each

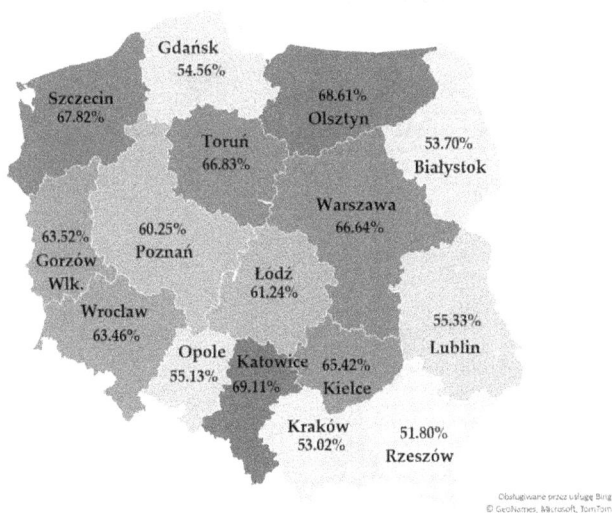

Figure 9.4 Share of mixed waste in the total waste in provincial cities in Poland in 2021 [in %].

Source: Own work using a map from Microsoft Excel based on information from the Local Data Bank.

of the assessed environmental categories, which makes it difficult to consider them fully smart in the issues analyzed in this section.

9.3 Environmental conditions in Polish Cities – survey perspective

This section presents the results of surveys relating to the environmental conditions of 287 Polish cities. Table 9.1 includes ratings for each factor along with the values of descriptive statistics assigned to each question.

The first two survey questions dealt with general perceptions of environmental issues. Thus, from the results shown in Table 9.1, it is clear that city authorities recognize the importance of environmental protection and highly rate the importance of this factor in shaping the quality of urban life. Both the value of the mean and the dominant and median values indicate that the environment is important or very important to cities.

Table 9.1 Basic statistics for assessing environmental conditions

Evaluated conditions	Arithmetic mean	Median	Mode	Standard deviation	Coefficient of variation
ENVIRONMENTAL					
Relevance of environmental quality	4.11	4.00	4.00	0.76	18.49%
The difficulty of improving environmental quality	2.72	3.00	2.00	1.03	37.87%
The level of environmental pollution in the city compared to the average in Poland	3.21	3.00	3.00	1.08	33.64%
Segregation of municipal waste in the city	3.51	4.00	4.00	0.98	27.92%
Percentage of residents using wastewater treatment plants compared to the average in Poland	3.72	4.00	4.00	1.04	27.96%
Impact of environmental organizations on entrepreneurship in the city	2.95	3.00	3.00	0.82	27.80%

Source: Own elaboration based on questionnaire surveys.

The surveyed entities are quite consistent in this regard, as the coefficient of variation for the analyzed response is at a very low level. Nevertheless, the majority of officials surveyed said that improving environmental quality is 'difficult'. This task is therefore a challenge for the surveyed cities, as confirmed by the statistical indicators analyzed earlier. They show that few cities are financially committed to environmental and climate protection, placing more priority on tasks assigned to municipalities. The financial barrier is therefore probably the most serious obstacle to achieving the environmental development goals of smart cities (Biernacka and Kronenberg, 2018; Żelazna et al., 2020). City governments recognize the importance of these goals, but are unable to take action for more effective and broader environmental and climate protection (Przywojska et al., 2019; Drożdż et al., 2021). This is a serious impediment to the full sustainability of Polish cities, since environmental and climate protection in exemplary fully developed smart cities is a prerequisite for maintaining their current status. Importantly, the analyzed cities vary quite significantly in their responses among themselves, as indicated by the value of the standard deviation and the coefficient of variation, which means that in the surveyed group, there are also entities that do not have difficulties with environmental protection, as well as those that perceive it as very difficult.

Despite the difficulties in implementing pro-environmental tasks, most of the surveyed cities rate the state of the city's environment as average compared to the country as a whole. This is not a satisfactory assessment, but in the context of the statistical data presented, it should be considered objective and illustrative of Polish problems with the greening of urban life. They are also evident on the scale of the economy as a whole, which has been struggling for many years to make the economic transition and move away from fossil fuels effective (Czarnecka et al., 2022; Kinelski, 2022). Thus, we can conclude that local and regional environmental problems are to some extent a derivative of national difficulties including the lack of coherent, consistently implemented environmental and energy policies (Gantowska and Moryń-Kucharczyk, 2019; Brauers and Oei, 2020; Pietrzak et al., 2021), which should not, however, be an excuse for downplaying the importance of the environment in shaping the quality of life of residents.

In the city's assessment of individual aspects of environmental protection, the issue of wastewater treatment fared quite well, with the majority of respondents rating it well. We should also add that the cities did not differ significantly in this regard, as documented by the measures of variability in Table 9.1. Such an outstanding rating is largely due to the European Union's financial support for projects implemented in Poland in the area of expanding the sewage network and increasing the availability of wastewater treatment plants (Piasecki, 2019; Kubiak-Wójcicka and Kielik, 2021). They have allowed many cities, especially smaller ones, to improve their infrastructure and move toward a closed-loop economy in this aspect of urban life (Rosiek, 2020).

The waste segregation process was rated slightly worse, at around 3.5, which means that Polish cities are aware that still too much of municipal waste is mixed waste. Thus, both the city authorities and the residents themselves still have a lot of work to do in this regard, which was also confirmed by the provincial statistical indicators presented in the previous section. The difficulties associated with the segregation and management of municipal waste are primarily related to the costliness of these processes and the lack of modern technological solutions to facilitate them (Mesjasz-Lech and Michelberger, 2019; Mesjasz-Lech, 2021; Jonek-Kowalska, 2022).

On the contrary, the impact of environmental organizations on entrepreneurship in the city was rated very poorly (average: below 3.0). This implies a lack of interaction between city authorities – business – environmental organizations and a failure to implement the guidelines of the fivefold economic helix. Thus, the surveyed cities, for the most part, have not reached the more advanced stages of development of smart urban structures, and in this context, it is difficult to consider them fully sustainable and smart (Wierzbicka, 2020).

In the next part of the discussion – following the pattern of previous chapters – the relationships of the size and wealth of the studied cities with the analyzed environmental conditions were identified. The results of this step are included in Table 9.2.

Table 9.2 Basic statistics for assessing environmental conditions

Evaluated conditions	Spearman's rank correlation coefficient	
	Population	Income level
ENVIRONMENTAL		
Relevance of environmental quality	0.2349**	0.0247
The difficulty of improving environmental quality	−0.0445	−0.0706
Segregation of municipal waste in the city	0.1036	0.0144
The level of environmental pollution in the city compared to the average in Poland	−0.0647	−0.0121
Percentage of residents using wastewater treatment plants compared to the average in Poland	0.4109**	0.2804**
Impact of environmental organizations on entrepreneurship in the city	0.1083	0.0618

Source: Own elaboration based on questionnaire surveys.
* $p < 0.05$; ** $p < 0.01$.

Statistically significant relationships were found in only three cases. The first relates to the weak positive linear correlation between the assessment of environmental relevance and city size, meaning that awareness of the importance of environmental aspects tends to be higher in larger cities than in entities with smaller populations. The next two cases refer to the percentage of residents using wastewater treatment plants against the average in Poland. In light of the results, this percentage is higher in larger and more affluent cities. However, this relationship, despite statistical significance, is moderate and weak, respectively.

Thus, the correlation analysis shows that environmental aspects in most cases do not depend directly on the size and economic situation of the city, and even if such a relationship exists, it is weak or moderate in nature. Such an inference should be seen as an opportunity for all surveyed cities to take effective environmental measures. Indeed, the problem of large cities may be limiting the effects of excessive urbanization, which will not be the affliction of smaller entities, which can thus use the sources of their environmental competitive advantage.

Summarizing the considerations carried out in this chapter, it should be said that environmental conditions and their role in improving the quality of life of residents are recognized by the authorities of the surveyed cities, but are not prioritized. This results in very low budget expenditures on environmental and climate protection, a fairly high share of mixed waste in total waste, and often a low or average share of green areas in the total area of cities.

As a result, there is not full sustainability in the surveyed cities. There is a focus on economic, technical-infrastructural and social goals, with the majority of the priorities listed being related to the basic living needs of residents and addressing issues such as road and housing infrastructure or social assistance (Sikora-Fernandez, 2018; Rześny-Cieplińska et al., 2021). In developing and emerging economies, the satisfaction of these needs is not yet sufficient, preventing both the economy as a whole and the cities operating within it from moving up the pyramid of economic and civilizational development. In the identified conditions, it will therefore be difficult to carve out smart cities that are fully sustainable and mature. This is also confirmed by the examples of the leaders of the smart rankings: Warsaw and Wrocław, where environmental statistical indicators did not fare well compared to the other provincial cities, which means that these units, too, are struggling with the implementation of pro-environmental tasks and are not fully effective and efficient in this context either.

It seems that little will change in the near future in the economic situation of Polish cities due to the crisis caused by the COVID-19 pandemic and the Russia-Ukraine conflict, which have devastated the state of public finances in Poland (Miciuła et al., 2021). Rising inflation and reduced state budget revenues have resulted in a reduction of funds for the implementation of all tasks entrusted to municipalities, so the already very low environmental spending, which is not a priority development direction, may decrease in the near future.

Unfortunately, an additional factor hindering the development of pro-environmental investments in Polish cities is the government's current political disputes with the European Union. Their consequence is the withholding of EU funding from the National Reconstruction Program, which makes it practically impossible to implement most of the development plans of Polish cities, which for many years have been financed precisely with EU funds.

Despite the above circumstances, efforts to promote the Smart City concept should not be halted, as it indicates a clear and desirable direction for urban development. In idealistic assumptions, it may seem utopian, difficult to achieve and unrealistic; however, it is important to strive for this ideal, including especially in environmental issues. After all, current climatic conditions emphatically expose the importance of the environment for the lives of present and future generations, and failure to act in this area could be catastrophic. The difficulty of improving the environment – accentuated in municipal responses – and insufficient financial resources cannot be a permanent obstacle to achieving sustainable development goals. The surveyed cities should therefore at least try not to deteriorate an existing urban ecosystem, which often accompanies progressive urbanization.

Bibliography

Anguluri, R., Narayanan, P. (2017). Role of green space in urban planning: Outlook towards smart cities. *Urban Forestry & Urban Greening, 25*, 58–65. https://doi.org/10.1016/j.ufug.2017.04.007

Artmann, M., Kohler, M., Meinel, G., Gan, J., Ioja, I.-C. (2019). How smart growth and green infrastructure can mutually support each other—A conceptual framework for compact and green cities. *Ecological Indicators, 96*(2), 10–22. https://doi.org/10.1016/j.ecolind.2017.07.001

Biernacka, M., Kronenberg, J. (2018). Classification of institutional barriers affecting the availability, accessibility and attractiveness of urban green spaces. *Urban Forestry & Urban Greening, 36*, 22–33. https://doi.org/10.1016/j.ufug.2018.09.007

Brauers, H., Oei, P.-Y. (2020). The political economy of coal in Poland: Drivers and barriers for a shift away from fossil fuels. *Energy Policy, 144*, 111621.

Cepeliauskaite, G., Keppner, B., Simkute, Z., Stasiskiene, Z., Leuser, L., Kalnina, I., Kotovica, N., Andiņš, J., Muiste, M. (2021). Smart-mobility services for climate mitigation in urban areas: Case studies of Baltic countries and Germany. *Sustainability, 13*, 4127. https://doi.org/10.3390/su13084127

Chekima, B., Chekima, S., Wafa, S.A., Wafa, S.K., Igau, O.A., Sondoh Jr., S.L. (2016). Sustainable consumption: The effects of knowledge, cultural values, environmental advertising, and demographics. *International Journal of*

Sustainable Development & World Ecology, 23(2), 210–220. https://doi.org/1 0.1080/13504509.2015.1114043

Chen, Ch., San, Y., Lan, Q., Jiang, F. (2020). Impacts of industrial agglomeration on pollution and ecological efficiency-A spatial econometric analysis based on a big panel dataset of China's 259 cities. *Journal of Cleaner Production, 258*, 120721. https://doi.org/10.1016/j.jclepro.2020.120721

Chodkowska-Miszczuk, J., Rogatka, K., Lewandowska, A. (2021). The Anthropocene and ecological awareness in Poland: The post-socialist view. *The Anthropocene Review*. https://doi.org/10.1177/20530196211051205

Czarnecka, M., Chudy–Laskowska, K., Kinelski, G., Lew, G., Sadowska, B., Wójcik-Jurkiewicz, M., Budka, B. (2022). Grants and funding for the processes of decarbonization in the scope of sustainability development—The case from Poland. *Energies, 15*, 7481. https://doi.org/10.3390/en15207481

Drożdż, W., Kinelski, G., Czarnecka, M., Wójcik-Jurkiewicz, M., Maroušková, A., Zych, G. (2021). Determinants of decarbonization—How to realize sustainable and low carbon cities? *Energies, 14*, 2640. https://doi.org/10.3390/en14092640

Erős, N., Török, Z., Hossu, C.-A., Réti, K.O., Maloş, C., Kecskés, P., Morariu, S.-D., Benedek, J., Hartel, T. (2022). Assessing the sustainability related concepts of urban development plans in Eastern Europe: A case study of Romania. *Sustainable Cities and Society, 85*, 104070. https://doi.org/10.1016/j.scs.2022.104070

Gherbi, M. (2012). Problematic of environment protection in Algerian cities. *Energy Procedia, 18*, 265–275. https://doi.org/10.1016/j.egypro.2012.05.038

Gantowska, R., Moryń-Kucharczyk, E. (2019). Current status of wind energy policy in Poland. *Renewable Energy, 135*, 232–237. https://doi.org/10.1016/j.renene.2018.12.015

Hasan, S.E. (2004). Public awareness is key to successful waste management. *Journal of Environmental Science and Health, Part A, 39*(2), 483–492. https://doi.org/10.1081/ESE-120027539

Jonek-Kowalska, I. (2022). Municipal waste management in Polish cities—Is it really smart? *Smart Cities, 5*, 1635–1654. https://doi.org/10.3390/smartcities5040083

Kalbarczyk, E., Kalbarczyk, R. (2022). Credibility assessment of municipal climate change adaptation plans using the ex-ante method: A case study of Poland. *Sustainable Cities and Society, 87*, 104242. https://doi.org/10.1016/j.scs.2022.104242

Kaluarachchi, Y. (2021). Potential advantages in combining smart and green infrastructure over silo approaches for future cities. *Frontiers of Engineering Management, 8*, 98–108. https://doi.org/10.1007/s42524-020-0136-y

Kinelski, G. (2022). Smart-city trends in the environment of sustainability as support for decarbonization processes. *Energy Policy Journal, 25*(2), 109–136. https://doi.org/10.33223/epj/149739

Kopacova, H. (2019). Reflexion of citizens' needs in city strategies: The case study of selected cities of Visegrad group countries. *Cities, 84*, 159–171. https://doi.org/10.1016/j.cities.2018.08.004

Kowalczyk, M. (2022). Environmental Protection Programmes in selected Polish communities as the first step towards sustainable development. *Zeszyty Teoretyczne Rachunkowości, 46*(2), 137–155.

Kronenberg, J., Haase, A., Łszkiewicz, E., Antal, A., Baravikova, A., Biernacka, M., Dushkova, D., Filčak, R., Haase, D., Ignatieva, M., Khmara, Y., Niță, M.R., Onose, D.A. (2020). Environmental justice in the context of urban green space availability, accessibility, and attractiveness in postsocialist cities. *Cities, 106,* 102862. https://doi.org/10.1016/j.cities.2020.102862

Kubiak-Wójcicka, K., Kielik, M. (2021). The state of water and sewage management in Poland. In: Zeleňáková, M., Kubiak-Wójcicka, K., Negm, A.M. (eds) *Quality of Water Resources in Poland.* Springer Water. Springer: Cham. https://doi.org/10.1007/978-3-030-64892-3_16

Lewandowska, A., Szymańska, D. (2021). Ecologisation of Polish cities in the light of selected parameters of sustainable development. *Sustainable Cities and Society, 64,* 102538. https://doi.org/10.1016/j.scs.2020.102538

Lin, J., Long, C., Yi, Ch. (2021). Has central environmental protection inspection improved air quality? Evidence from 291 Chinese cities. *Environmental Impact Assessment Review, 90,* 106621. https://doi.org/10.1016/j.eiar.2021.106621

Mesjasz-Lech, A. (2021). Municipal urban waste management—Challenges for Polish cities in an era of circular resource management. *Resources, 10,* 55. https://doi.org/10.3390/resources10060055

Mesjasz-Lech, A., Michelberger, P. (2019). Sustainable waste logistics and the development of trade in recyclable raw materials in Poland and Hungary. *Sustainability, 11,* 4159. https://doi.org/10.3390/su11154159

Miciuła, I., Tylżanowski, R., Rogowska, K., Stępień, P. (2021). Managing economic and social changes during the COVID-19 pandemic: A case study for Szczecin and Krakow. *European Research Studies Journal, XXIV*(4), 365–376.

Natanian, J., Auer, Th. (2020). Beyond nearly zero energy urban design: A holistic microclimatic energy and environmental quality evaluation workflow. *Sustainable Cities and Society, 56,* 102094. https://doi.org/10.1016/j.scs.2020.102094

Never, B., Albert, J.R.G. (2021). Unmasking the middle class in the Philippines: Aspirations, lifestyles and prospects for sustainable consumption. *Asian Studies Review, 45*(4), 594–614. https://doi.org/10.1080/10357823.2021.1912709

Niedźwiecka-Filipiak, I., Potyrała, J., Filipiak, P. (2015). Contemporary management of green infrastructure within the borders of Wrocław functional area (WrOF). *Architektura Krajobrazu, 2,* 4–27.

Niedźwiecka-Filipiak, I., Rubaszek, J., Potyrała, J., Filipiak, P. (2019). The method of planning green infrastructure system with the use of landscape-functional units (method LaFU) and its implementation in the Wrocław functional area (Poland). *Sustainability, 11,* 394. https://doi.org/10.3390/su11020394

Piasecki, A. (2019). Water and sewage management issues in rural Poland. *Water, 11,* 625. https://doi.org/10.3390/w11030625

Pietrzak, M.B., Igliński, B., Kujawski, W., Iwański, P. (2021). Energy transition in Poland—Assessment of the renewable energy sector. *Energies, 14,* 2046. https://doi.org/10.3390/en14082046

Przywojska, J., Podgórniak-Krzykacz, A., Wiktorowicz, J. (2019). Perceptions of priority policy areas and interventions for urban sustainability in Polish municipalities: Can Polish cities become smart, inclusive and green? *Sustainability, 11,* 3962. https://doi.org/10.3390/su11143962

Portal internetowy miasta Białystok. https://www.bialystok.pl, [access data: 11.11.2022].

Rosiek, K. (2020). Directions and challenges in the management of municipal sewage sludge in Poland in the context of the circular economy. *Sustainability, 12*, 3686.

Rześny-Cieplińska, J., Szmelter-Jarosz, A., Moslem, S. (2021). Priority-based stakeholders analysis in the view of sustainable city logistics: Evidence for Tricity, Poland. *Sustainable Cities and Society, 67*, 102751. https://doi.org/10.1016/j.scs.2021.102751

Satoła, Ł., Milewska, A. (2022). The concept of a smart village as an innovative way of implementing public tasks in the era of instability on the energy market—Examples from Poland. *Energies, 15*, 5175. https://doi.org/10.3390/en15145175

Serbanica, C., Constantin, D.-L. (2017). Sustainable cities in central and eastern European countries. Moving towards smart specialization. *Habitat International, 68*, 55–63. https://doi.org/10.1016/j.habitatint.2017.03.005

Sikora-Fernandez, D. (2018). Smarter cities in post-socialist country: Example of Poland. *Cities, 78*, 52–59. https://doi.org/10.1016/j.cities.2018.03.011

Ustawa z dnia 5 czerwca 1998 o samorządzie województwa (Dz. U. 1998 Nr 91 poz. 576 z późn. zm.).

Ustawa z dnia 8 marca 1990 roku o samorządzie gminnym (Dz. U. 1990 Nr 16 poz. 95 z późn. zm.)

Wang, Q., Zhao, Z., Shen, N., Liu, T. (2015). Have Chinese cities achieved the win–win between environmental protection and economic development? From the perspective of environmental efficiency. *Ecological Indicators, 51*, 151–158. https://doi.org/10.1016/j.ecolind.2014.07.022

Wierzbicka, W. (2020). Socio-economic potential of cities belonging to the Polish National Cittaslow Network. *Oeconomia Copernicana, 11*(1), 203–224. https://doi.org/10.24136/oc.2020.009

Witkowski, J. (2019). The engagement of commune self-governments in the functioning of legal forms of environmental protection by the example of selected communes in the Lubelskie Voivodship. *Zeszyty Naukowe. Organizacja i Zarządzanie. Politechnika Śląska, 137*, 221–231.

Zawilińska, B., Brańka, P., Majewski, K., Semczuk, M. (2021). National parks—Areas of economic development or stagnation? Evidence from Poland. *Sustainability, 13*, 11351. https://doi.org/10.3390/su132011351

Żelazna, A., Bojar, M., Bojar, E. (2020). Corporate social responsibility towards the environment in Lublin Region, Poland: A comparative study of 2009 and 2019. *Sustainability, 12*, 4463.

10 Directions for improving the quality of life in Central and Eastern Europe cities

10.1 Barriers to improving the quality of life in CEE smart cities

Based on the considerations and studies conducted, it can be generally concluded that smart cities in Central and Eastern Europe are familiar with the Smart City concept, are interested in its implementation and are making successful attempts to use smart city solutions (Dohn et al., 2022; Kramarz et al., 2022). Nevertheless, cities in the region have a lower level of maturity in terms of being smart than units located in developed European economies. The main reason for this is their relatively short period of operation under free market conditions. This is because the countries of Central and Eastern Europe were for many years under the political and economic influence of the Union of Soviet Socialist Republics and only in the 1990s did they begin the process of economic transformation (Esses et al., 2021; Pakulska, 2021). For them, catching up with the more developed countries of the European Union is a very difficult and complex challenge that will certainly take many more years to complete (Varga, 2020).

Details on the development of smart cities in Central and Eastern Europe developed from previous chapters of this monograph are included in Table 10.1, which contains the weaknesses and strengths of the studied urban agglomerations.

The summary of strengths and weaknesses of cities operating in Central and Eastern Europe in Table 10.1 shows that they have far more imperfections than strengths. Among the promising aspects of Central and Eastern European cities, one can certainly point to mobility and transportation, dynamically developing capitals and a growing interest in implementing Smart City solutions. On the contrary, aspects that significantly degrade the quality of life of residents are the inaccessibility of modern housing infrastructure, lack of care for the environment and aging urban communities.

DOI: 10.4324/9781003358190-10

Table 10.1 Strengths and weaknesses of smart cities in Poland and other Central and Eastern European (CEE) countries

Rated area	Weak points	Strong points
Economic and financial	CEE Dominance of capitals in terms of development, innovation and entrepreneurship – unbalanced against other cities. Lack of strategies and development plans for smart cities. Centralization of decisions and lack of legislation to support the development of smart city solutions. Weak development of cities in Eastern Poland. The need to meet the challenges of digitization and economic and energy transformation. Low wages and few jobs in high-growth sectors. POLAND Large variations in the level of budget revenues. Low share of property (infrastructure) expenditures in total expenditures. Financing capital expenditures from the budget deficit resulting in debt accumulation. Lack of activating and motivational forms of support for entrepreneurs, as well as the unemployed.	CEE Knowledge of Smart City concepts. Implementation of Smart City solutions and projects. Use of EU funds for SC investments. Seeking SC's own development paths tailored to the specific characteristics of CEE countries. Large number of SC studies in Romania. Well-regarded degree of development of Czech Smart Cities. POLAND A good assessment of the city's financial stability and general investment conditions. Promoting the city as an investor-friendly destination. Slightly better than average coverage of financial support for the unemployed.
Technological	CEE Low – compared to developed economies – access to IT and ICT technologies. Fear of unwanted use of modern technology (data security; privacy protection). POLAND Low levels of entrepreneurship and innovation. Poor access to business incubators and technology parks. Lack of support from the city for the development of small and medium-sized enterprises.	CEE Well-functioning and numerous startups in cities considered smart. Associating SC with artificial intelligence and machine learning. Willingness to use modern technology to protect urban health and life. POLAND Good accessibility of IT infrastructure for residents. City offices are well equipped with modern IT equipment. Good ability of the city government to use digital communication with residents.

(*Continued*)

Table 10.1 (Continued)

Rated area	Weak points	Strong points
Infrastructural	CEE Poor quality and availability of housing infrastructure severely compromising quality of life. Development of new, exclusive, and gated housing developments around cities disrupting their sustainability. POLAND Very low availability of housing infrastructure (old and newly built). Less accessible rail and air transportation in smaller cities.	CEE A large number of implementations of smart solutions in urban mobility and transport. Recognizing the need to include the community in the creation of intelligent mobility. POLAND Well-developed public transportation network and road network. Incurring significant budget expenditures on transportation and communications.
Demographic	CEE Progressive aging of urban communities. High risk of social, economic and digital exclusion. POLAND Rapid and progressive aging of urban communities. Extremely low (negative) birth rate.	CEE – POLAND Slower community aging and better birth rates in cities with high reputations and very good levels of economic development.
Educational	CEE Digital competency imbalance resulting in exclusion. POLAND Poor access to university education in smaller cities. Low availability of training to improve competencies, including IT.	CEE Offering modern public services to residents. POLAND Guaranteed primary and secondary education at a very good and good level.
Social (participatory)	CEE The existence of a social digital divide between cities. Excessive neoliberalization resulting in marginalization of social issues. Little interest in public participation. POLAND Relatively weak development of central and local democracy. Reluctance of local governments to promote public participation. Lack of interest and use of the sharing economy.	CEE Recognizing the importance of public participation in shaping the quality of urban life. POLAND Growing interest in public participation among residents. Sanctioning forms of participation in local government regulations. Increasing numbers of public participation initiatives.

(*Continued*)

Table 10.1 (Continued)

Rated area	Weak points	Strong points
Environmental	CEE Relatively low environmental awareness of residents. Low involvement of the city government in environmental issues related to tolerance of social inequality, lack of solidarity in society, lack of responsibility for the public interest, extreme individualization and disregard for social interests. POLAND A very low level of per capita environmental and climate spending indicating the low importance of environmental issues. Relatively high share of non-segregated waste in total waste in most cities surveyed. Seeing the issue of improving environmental quality as a very difficult task. Low level of involvement of environmental organizations in the life of the city.	CEE Solidarity aid from more developed EU countries. Systematic but slow growth of municipal interest in environmental protection. POLAND Recognizing the importance of environmental protection in shaping the quality of life. A good level of urban wastewater treatment facilities.

Source: Own study.

The research conducted in the monograph also provides information on the relationship between the various factors shaping the quality of life of residents and the size and wealth of the cities studied. The information is shown in Table 10.2.

The information in Table 10.2 shows that many of the determinants shaping residents' quality of life are positively correlated with city size. This means that people live better in large cities. This is due to large cities' offering high availability of transportation, housing, and investment infrastructure. They also provide better educational opportunities.

Far fewer relationships were identified between city wealth and the analyzed determinants of urban quality of life. These correlations relate primarily to economic and financial determinants, which seems to be a natural consequence of their nature. In addition, wealthier cities have better housing infrastructure and higher education offerings.

The above conclusions allow us to conclude that smart cities are primarily large cities, so in the future we can expect further intensive development of these cities and their progressive urbanization. They will attract more residents, which may worsen the imbalance between urban

Table 10.2 Identified relationships between wealth and city size and quality of life in each area of the analysis conducted

Dependence type	Dependence description
Dependence of conditions on city size	Statistically significant, but weak and moderate relationship of almost all (except debt) economic and financial parameters. Strongest for the ability to obtain investment support from city funds and overall employment opportunities in the city.
	A moderate relationship between entrepreneurship and innovation.
	Significant dependence on the availability of housing infrastructure.
	Average dependence of public transportation and rail accessibility.
	Average and weak positive correlation of availability of primary, secondary and higher education and in-service training.
	Low positive correlation between sharing economy development.
	All the educational determinants listed in the survey questionnaire are positively correlated with the size of the city expressed in terms of population.
	The level of demographic, social and educational determinants is more strongly related to the size of a city than to its wealth.
	Moderate relationship between city size and initiatives and organizational and financial support for sharing economy.
	The percentage of residents using wastewater treatment plants compared to the average in Poland is higher in wealthier cities.
	Weak positive linear correlation between ratings of environmental significance and city size.
	The percentage of residents using wastewater treatment plants compared to the average in Poland is higher in larger cities.
Dependence of conditions on city financial health	Statistically significant, but weak and moderate relationship for almost all (except debt and unemployment welfare) economic and financial parameters. Strongest for stability of the city's economic situation and overall employment opportunities in the city.
	Moderate dependence on the availability of housing infrastructure.
	Low positive correlation between the availability of higher education and the possibility of receiving in-service training.

Source: Own study.

and rural areas. In the context of the analysis carried out, such a trend is also a serious threat to the natural environment of cities in Central and Eastern Europe. It is a very serious threat due to the low interest in environmental problems of both authorities and residents.

10.2 Recommendations for improving the quality of life in smart cities in Central and Eastern Europe

Based on the results obtained, recommendations for improving the quality of life in CEE cities were developed. They are included in Table 10.3, based on the different areas of assessment.

The results obtained and the recommendations made make it possible to conclude that the surveyed cities aspiring to be smart show the greatest imbalance in the environmental area, to which both city authorities

Table 10.3 Recommendations for improving the quality of life in the surveyed cities, considering the areas analyzed

Rated area	Recommendations
Economic and financial	Developing good quality and comprehensive Smart City development strategies.
	Seeking external and non-refundable sources of financing for infrastructure investments.
	Monitoring debt levels to avoid excessive budget burdens in the long term.
	Encouraging investors to start businesses not only through promotional activities but also through administrative, organizational and financial support.
	Design and implementation of activation and motivational forms of support for the unemployed.
	More active involvement in creating entrepreneurship and innovation in the city, even at the expense of reducing passive forms of support, such as monetary; unconditional social assistance.
Technological	Acquisition and development of IT and ICT technologies.
	Strengthening digital competence among city government employees and residents.
	Supporting local entrepreneurship, especially small and medium-sized enterprises.
	Promoting startups and providing them with institutional and financial assistance.
	Development of business incubators and technology parks, also in smaller cities.
Infrastructural	Intensive development of municipal housing infrastructure.
	Expanding the rail network to improve the quality and accessibility of this form of transportation.

(Continued)

Table 10.3 (Continued)

Rated area	Recommendations
Demographic	Monitoring the state of the aging population and taking measures to halt the process.
	Caring for seniors and preventing their social, economic and technological exclusion.
	Undertaking initiatives and local programs to improve birth rates.
Educational	Maintaining and improving the level of primary and secondary education.
	Facilitating access to regional university centers.
	Scaling up lifelong learning activities, including raising digital competencies of local communities.
Social (participatory)	Preventing social, economic and digital exclusion.
	Increasing local government involvement in participatory initiatives.
	Expanding participatory offerings beyond the limits set by local government laws.
	Promoting the sharing economy as part of sustainable consumption and environmental protection.
Environmental	Increasing municipal commitment to environmental and climate protection.
	Increasing spending on environmental and climate protection.
	Increasing the share of green areas in the total area of cities, especially those heavily industrialized.
	Increasing the share of segregated waste in total waste in all surveyed cities.
	Seeking budgetary resources and external financing for the implementation of environmental investments.
	Raising awareness among city residents of the need to take environmentally friendly measures.
	Inclusion of environmental organizations in the life of the city.

Source: Own study.

and residents contribute. Thus, much remains to be done in this area in terms of education, awareness and reality. The situation is slightly better in the social area. Nevertheless, there is still a lack of active involvement of citizens in city management and in shaping the quality of urban life. The technological, mobility and economic areas should be considered fairly well developed, which alludes to the origins of the Smart City concept.

Bibliography

Dohn, K., Kramarz, M., Przybylska, E. (2022). Interaction with city logistics stakeholders as a factor of the development of Polish cities on the way to becoming smart cities. *Energies, 15*, 4103. https://doi.org/10.3390/en15114103

Esses, D., Csete, M.S., Németh, B. (2021). Sustainability and digital transformation in the Visegrad Group of Central European countries. *Sustainability, 13,* 5833. https://doi.org/10.3390/su13115833

Kramarz, M., Dohn, K., Przybylska, E., Jonek-Kowalska, I. (2022). Smart city: A holistic approach. In: *Urban Logistics in a Digital World.* Palgrave Macmillan: Cham. https://doi.org/10.1007/978-3-031-12891-2_1

Pakulska, T. (2021). Green energy in Central and Eastern European (CEE) countries: New challenges on the path to sustainable development. *Energies, 14,* 884. https://doi.org/10.3390/en14040884

Varga, M. (2020). The return of economic nationalism to East Central Europe: Right-wing intellectual milieus and anti-liberal resentment. *Nations and Nationalism, 27*(1), 206–222. https://doi.org/10.1111/nana.12660

11 Conceptualizing research findings on quality of life in smart cities

11.1 Characteristics of the triad: sustainable development – Smart City – the quality of life

When analyzing the relationships between the triad-forming concepts discussed in this monograph, namely, quality of life (Dawood, 2019; Cai et al., 2021), sustainable development (Bibri and Krogstie, 2017) and Smart City (Giffinger, 2010; Ryba, 2017), we can state that the concepts of quality of life and sustainable development largely overlap, with the concept of sustainable development being more focused on the long-term and holistic approach and the concept of quality of life focusing more on the short-term and satisfaction of residents; other aspects, for example, the natural environment, are taken into account only when they contribute to the improvement of the quality of life. For the concept of quality of life to be also used to analyze the long-term functioning of the city, issues related to sustainable development should be taken into account to a greater extent (Shen et al., 2018; Abunazel et al., 2019). Including sustainable development in the concept of quality of life allows a more holistic view of the quality of life and also takes into account the long-term perspective, i.e., paying attention not only to the level of the current quality of life of residents but also to the ability of the city governments to maintain or improve the quality of life of its residents in the long term.

The relationships between the concepts described above are shown in Figure 11.1. Implementing the Smart City concept affects both the quality of life and sustainable development. Including sustainable development in the quality of life analysis leads to the emergence of the holistic concept of quality of life.

Considering the quality of life in this holistic way, the following areas of quality of life can be distinguished, illustrated in Figure 11.2:

- economic and financial;
- technological;

DOI: 10.4324/9781003358190-11

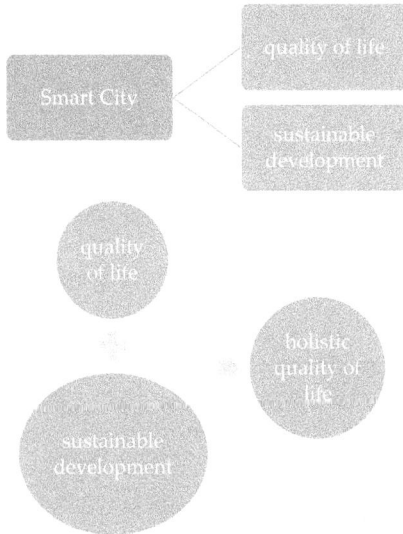

Figure 11.1 Relations between Smart City, quality of life and sustainable development.

Source: Own study.

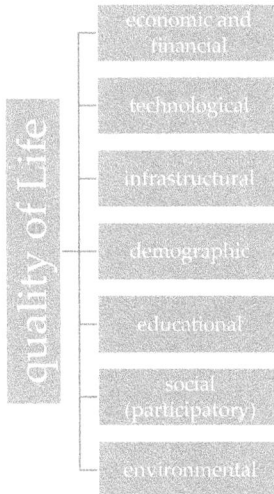

Figure 11.2 Areas of a holistic model of quality of life.

Source: Own study.

- infrastructural;
- demographic;
- educational;
- social (participatory);
- environmental.

When implementing the Smart City concept, the goal is to improve residents' quality of life and implement sustainable development in the city (Ramirez Lopez and Grijalba Castro, 2021). Properly implemented Smart City activities should contribute to improving the quality of life of residents (Chen, 2010; McFarlane and Söderström, 2017; Cardullo and Kitchin, 2019), while problems in this area, such as forgetting about digitally excluded people and deepening their exclusion as a result of the implementation of modern digital city services, will contribute to a decrease in the quality of life of residents.

11.2   A model of pro-quality management of a smart sustainable city

Based on the results obtained, recommendations for improving the quality of life in CEE cities were developed. They are included in Table 10.3, based on the different areas of assessment.

The implementation of the Smart City concept in the city can lead to both an improvement in the quality of life and, in some situations, a decrease in the quality of life. Table 11.1 presents both negative and positive consequences for the quality of life, which in individual areas may arise as a result of the implementation of the Smart City concept in the city.

Data collected in Table 11.1 show that the potentially positive impact of Smart City implementation on residents' quality of life definitely outweighs the negative ones. The implementation of Smart City solutions

Table 11.1 Analysis of the positive and negative impact of the Smart City implementation on individual areas of the quality of life in the city

Rated area	Positive impact	Negative impact
Economic and financial	Improvement of the city's economic development, increase in the inhabitants' wealth. Lower unemployment, finding a job is easier. Modern solutions reduce the costs of city management. Increased R&D spending in the city. Improved productivity.	Growing economic inequality, the rich getting richer and the poor getting poorer.

(*Continued*)

Table 11.1 (Continued)

Rated area	Positive impact	Negative impact
Technological	The city is becoming a hub of new technologies and creative solutions. Improved level of entrepreneurship and innovation of the inhabitants. The city attracts new investors and startups. Sustainable management of raw materials.	People who are not proficient in using modern technologies and those who cannot afford to finance them are excluded from the use of new technologies.
Infrastructural	Improved quality of ICT infrastructure. Improved quality of the built housing infrastructure – smart buildings. Better accessibility of transport, in particular public transport.	Growing density of population and infrastructure, which can be a threat to the environment.
Demographic	Opportunity to attract young, creative people from the country and the world. Slower aging of the population, higher birth rate.	Distortion of the demographic structure and the effect of crowding out the elderly.
Educational	Increased level of education, modern education educating advanced specialists.	Growing inequalities among children – unbalanced digital competences.
Social (participatory)	Opportunity for residents to participate in city management – participatory budgets, making decisions in a popular vote of residents. Improved functioning of democratic mechanisms in society. Easier access for citizens to city data. Smart healthcare improving the quality of medical services.	Exclusion from participation of residents who are unable to efficiently use modern technologies, e.g., a smartphone.
Environmental	Saving energy – smart grid. Promoting the sharing economy. Using the application to improve waste segregation. Pro-ecological smart transport – scooters and city bikes. Better involvement of environmental organizations in the life of the city. Use of clean energy. Reduced water consumption.	Part of the investment in new technologies may have a negative impact on the environment.

Source: Own study.

can have a positive impact on all analyzed areas of the quality of life model. The main problem is related to the issues of digital exclusion, in particular older people who have difficulties in using new technologies and other types of exclusion, namely, younger people, with lower incomes, with disabilities, lower education, immigrants, etc., who may have problems with the use of new technological solutions also due to the lack of funds for the purchase of appropriate equipment or the language barrier. Avoiding a situation where the implementation of Smart City solutions will lead to a decrease in the quality of life of residents is conditioned by solving the problem of digital and economic exclusion and pursuing a policy of including all residents in the new solutions as broadly as possible.

When analyzing the relationships between the city's scope of Smart City implementation and the quality of life, the matrix model presented in Figure 11.3 can be used.

The model was developed taking into account two variables of the discussed triad, i.e., Smart City and quality of life. From this point of view, we can distinguish four types of cities:

- **Pro-quality inclusive Smart City** – cities that combine the use of modern technologies with an increase in the quality of life of all residents in a sustainable way. They take care of the uniform improvement of all indicators related to the quality of life, such as economic, technological, infrastructural, social or environmental. An important element of this type of city is a high degree of inclusivity, where care is taken not to leave any group of residents excluded (due to age, ethnicity or other factors). The city governments make consistent efforts to ensure that all its residents use modern smart technological solutions. In such cities, the implementation of the Smart City concept strongly contributes to improving the quality of life of residents. Examples are the world's leading Smart Cities, broad participation of citizens and caring for groups of potentially excluded people.
- **Smart city of inequalities** – cities strongly involved in the implementation of Smart City projects, in which very high inequalities characterize the level of participation of residents in these solutions. Some residents have access to Smart City solutions and use them very actively, while at the same time, there are numerous groups of residents who are excluded and may therefore feel isolated. In this group of cities, the implementation of smart solutions contributes to the improvement of the quality of life of some residents, but there are groups of people in the case of which they may lead to an increase in exclusion and a decrease in their quality of life. This includes large cities, including, for example, the capitals of Central and Eastern Europe.
- **Traditional city** – well-managed cities that create good living conditions for residents, but only to a small extent, implement Smart City concepts. In many countries, especially smaller cities are in this group, as they are involved in the implementation of smart solutions to a

much lesser extent than large cities. Despite this, many of them have a high level of quality of life. However, in the future, when technology becomes more widespread, it will be increasingly difficult to maintain a sufficient level of quality of life without Smart City solutions. Without the implementation of a smart grid for energy management, smart medical and tourist applications, etc., the quality of life of residents may decrease in the future. In the future, these types of cities will evolve either toward the pro-quality and inclusive Smart City model (path 1) or the smart city of inequality (path 2). City governments should take this into account in their strategy and try to implement Smart City solutions in the future in such a way as not to reduce the quality of life of residents and follow path 1. If the city does not implement Smart City solutions, it is at risk of technological and then economic backwardness, which may lead to path 3 – a slow 'slide' of the city toward the underdeveloped city model. This includes smaller cities from Central and Eastern Europe, especially those located in tourist areas.

- **Underdeveloped city** – cities usually located in countries/regions with a low degree of economic development, with a low level of quality of life for residents and low implementation of Smart City solutions. These types of cities will become more and more backward over time and will be a worse place to live with social inequalities and areas of poverty. This includes cities from countries with a very low level of economic development.

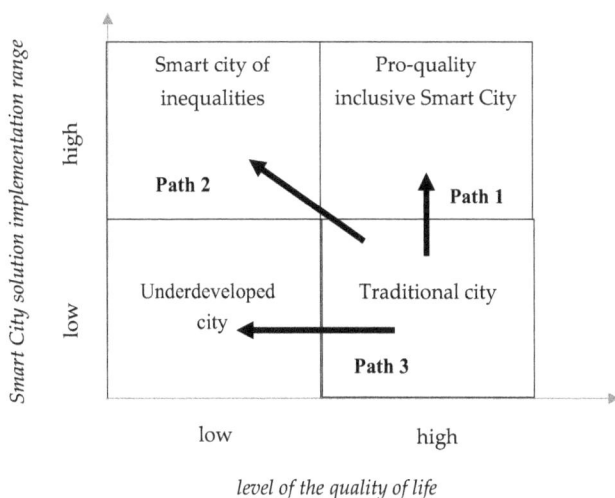

Figure 11.3 Matrix model of the relationship between the quality of life in the city and the scope of implementation of Smart City solutions.

Source: Own study.

The presented matrix model shows that in the long term, there is no alternative to implementing Smart City solutions in cities if you want to maintain a high quality of life there. A city not involved in the implementation of Smart City solutions may, as it has been described, follow one of the three development paths – the most favorable is path 1, the least favorable is path 3, i.e., complete abandonment of the implementation of modern technologies for city management.

In order for cities to follow path 1 and implement pro-quality management of Smart City implementation, they should implement the actions presented in the model shown in Figure 11.4.

Figure 11.4 Model of pro-quality management of the smart sustainable city.

Source: Own study.

Implementing the activities presented in the model should contribute to an increase in the positive impact of the implementation of smart city solutions on residents' quality of life, both in the short term and long term. In particular, in the case of Central and Eastern European countries, where the current level of implementation of Smart City solutions and the level of quality of life are moderate, the implementation of new technologies must become a factor that will improve the quality of life and allow cities in these countries to match the quality of life to leading cities worldwide.

Bibliography

Abunazel, A., Hammad, Y., Abd AlAziz, M., Gouda, E., Anis, W. (2019). Quality of life indicators in sustainable urban areas. *Academic Research Community Publication, 3*, 78.

Bibri, S.E., Krogstie, J. (2017). Smart sustainable cities of the future: An extensive interdisciplinary literature review. *Sustainable Cities and Society, 31*, 183–212. https://doi.org/10.1016/j.scs.2017.02.016

Cai, T., Verze, P., Truls, E., Johansen, B. (2021). The quality of life definition: Where are we going? *Uro Journal of Urology, 1*, 14–22. htps://doi.org/10.3390/uro1010003

Cardullo, P., Kitchin, R. (2019). Smart urbanism and smart citizenship: The neoliberal logic of 'citizen-focused' smart cities in Europe. *Environment and Planning C: Politics and Space, 37*(5), 813–830.

Chen, T.M. (2010). Smart grids, smart cities need better networks [editor's note]. *IEEE Network, 24*(2), 2–3.

Dawood, S.R.S. (2019). Sustainability, quality of life and challenges in an emerging city region of George Town, Malaysia. *Journal of Sustainable Development, 12*, 35.

Dohn, K., Kramarz, M., Przybylska, E. (2022). Interaction with city logistics stakeholders as a factor of the development of Polish cities on the way to becoming smart cities. *Energies, 15*, 4103. https://doi.org/10.3390/en15114103

Giffinger, R. (2010). Smart cities ranking: An effective instrument for the positioning of the cities? *ACE: Architecture, City and Environment, 12*, 7–15.

McFarlane, C., Söderström, O. (2017). On alternative smart cities: From a technology-intensive to a knowledge-intensive smart urbanism. *City, 21*, 312–328. htps://doi.org/10.1080/13604813.2017.1327166

Mohanty, S.P., Choppali, U., Kougianos, E. (2016). Everything you wanted to know about smart cities. *IEEE Consumer Electronics Magazine, 5*(3), 60–70.

Ramirez Lopez, L.J., Grijalba Castro, A.I. (2021). Sustainability and resilience in smart city planning: A review. *Sustainability, 13*, 181.

Ryba, M. (2017). What is a "Smart City" concept and how we should cal lit in polish, research. *Papers of Wrocław University of Economics, 467*, 82–90.

Shen, L., Huang, Z., Wong, S.W., Liao, S., Lou, Y. (2018). A holistic evaluation of Smart City performance in the context of China. *Journal of Cleaner Production, 200*, 667–679.

Summary

As part of the summary of this monograph, we have answered the research problems stated in the introduction. Next, reference was made to the postulated directions of development of smart cities in Central and Eastern Europe and the directions of further research on this topic.

It follows from the conducted literature studies that modern smart cities should be notable not only by a high degree of advancement of the implemented IT and ICT technologies but also by intensive development in the social and environmental areas. This means the inclusion of the local community and environmental organizations in the process of making current and strategic decisions and shaping the desired quality of life. Only then can we talk about the creation and development of smart and sustainable cities. The pursuit of sustainability is also associated with the successive achievement of development maturity of smart cities, which currently include generations from 1.0 to 4.0.

The Smart City concept and its implementation in modern smart cities raise a lot of controversy. Its opponents claim that on various levels, it deepens exclusion, fuels unlimited and selfish consumerism and devastates the natural environment. Nevertheless, regardless of the validity of this argument, it will be difficult to stop humanity's desire to improve the quality of life, which can only take place in smart and sustainable cities. It is therefore important to eliminate the shortcomings of the Smart City concept and to monitor the effects of its use.

In Central and Eastern Europe, due to the political and systemic past, the development of smart cities is slower than in developed European countries. Therefore, they can be considered generation 1.0 or 2.0 cities. They are certainly not fully balanced either. Thanks to the financial support of the European Union, these cities can engage in the development of transport and road infrastructure, which is one of their main achievements in terms of being smart. They also make successful attempts to implement IT and ICT solutions. Nevertheless, they are not holistic in nature. Rather, these are individual investment projects constituting a component to shape a favorable image of the city. A comprehensive

DOI: 10.4324/9781003358190-12

approach to creating a Smart City in this region is also hindered by the lack of uniform, good-quality strategic documents that would allow systematic and consistent, but also coherent, implementation of individual smart city solutions.

In the cities of Central and Eastern Europe, little attention is paid to social and environmental issues. Most likely, the reasons for this state of affairs are the ongoing needs of the inhabitants that have not yet been fully met and the rather strong neoliberalization resulting in a disregard for collective needs and lack of involvement in shaping the quality of life of future generations. As a result, urban communities are reluctant to engage in participatory and environmental initiatives. The city authorities also do not always encourage such initiatives. As a result, full and sometimes even partial sustainability of the studied cities is still not possible.

In the context of the above, the real social challenge for Central and Eastern European cities is to prevent problems related to the aging of the population and to provide current and future city dwellers with modern and ecological housing infrastructure. Participation-related issues then recede into the background, but there are no formal obstacles for them to be implemented in parallel. The more so that identifying the needs of the elderly, people with disabilities or the sick requires their involvement in order to look for the most effective ways to improve the quality of their lives.

Environmental challenges facing the cities of Central and Eastern Europe are even more numerous. They primarily concern the energy transformation and the use of renewable energy sources. It is also important to implement as many smart urban solutions as possible in the circular economy, especially including the management of urban waste. However, the implementation of these challenges will not be fully possible if both the city authorities and the urban community do not become aware of the importance of environmental protection and pro-ecological behavior in shaping the quality of life of current and future generations of residents. For these reasons, it is important to popularize and imitate good practices of developed economies and educate and systematically raise environmental awareness.

Concerning the quality of life in Polish cities surveyed, it can be said to be quite diverse. Large cities rate better in this respect as residents there have better access to education, training and transport infrastructure. They have more startups operating with entrepreneurship and innovation developing better and faster. Therefore, large cities attract investors more easily, and residents gain better access to job opportunities. Also, social initiatives are easier to implement there.

Development disproportions, which indicate certain unsustainability of Polish cities, are also noticeable in the context of the results of the analysis of 16 provincial cities. Cities located in western Poland are in a better financial situation, which may predispose them to faster and more effective implementation of smart city solutions. Nevertheless, the example of Lublin (a city in eastern Poland, often classified as smart in international terms) shows that much depends on the will and commitment of the city authorities. Their determination may cause that also in economically less favored regions smart urban solutions are implemented and systematically improve the quality of life of the inhabitants.

At the same time, it is noteworthy that the surveys show that only the economic conditions of urban life are directly related to the wealth of the city. Infrastructural, technological, social and environmental factors are no longer as numerous and strongly determined as before by the budget income of cities. This allows us to assume that smart city solutions are also within reach of cities with fewer financial resources.

This monograph presents a holistic assessment of the quality of life in cities aspiring to be smart in Central and Eastern Europe, with a strong emphasis on Poland. Further analyses should undoubtedly focus on the problematic issues related to the unsustainability of the studied cities. These include housing infrastructure, problems of social and economic exclusion, issues related to social participation and all issues related to environmental and climate protection.

Index

For Product Safety Concerns and Information please contact our EU
representative GPSR@taylorandfrancis.com
Taylor & Francis Verlag GmbH, Kaufingerstraße 24, 80331 München, Germany

www.ingramcontent.com/pod-product-compliance
Lightning Source LLC
Chambersburg PA
CBHW070713220326
41598CB00024BA/3135